Lecture Notes in Economics and Mathematical Systems

518

Founding Editors:

M. Beckmann
H. P. Künzi

Managing Editors:

Prof. Dr. G. Fandel
Fachbereich Wirtschaftswissenschaften
Fernuniversität Hagen
Feithstr. 140/AVZ II, 58084 Hagen, Germany

Prof. Dr. W. Trockel
Institut für Mathematische Wirtschaftsforschung (IMW)
Universität Bielefeld
Universitätsstr. 25, 33615 Bielefeld, Germany

Co-Editor:

C. D. Aliprantis

Editorial Board:

A. Basile, A. Drexl, G. Feichtinger, W. Güth, K. Inderfurth, P. Korhonen,
W. Kürsten, U. Schittko, P. Schönfeld, R. Selten, R. Steuer, F. Vega-Redondo

T0181904

Springer
Berlin
Heidelberg
New York
Barcelona
Hong Kong
London
Milan
Paris
Tokyo

Stefan Napel

Bilateral Bargaining

Theory and Applications

 Springer

Author

Stefan Napel
Institute of Economics Theory and Operations Research
University of Karlsruhe
Kaiserstraße 12
76128 Karlsruhe
Germany

Cataloging-in-Publication data applied for

Die Deutsche Bibliothek - CIP-Einheitsaufnahme

Napel, Stefan:
Bilateral bargaining : theory and applications / Stefan Napel. - Berlin ;
Heidelberg ; New York ; Barcelona ; Hong Kong ; London ; Milan ; Paris ;
Tokyo : Springer, 2002
 (Lecture notes in economics and mathematical systems ; 518)
 ISBN 3-540-43335-X

ISSN 0075-8450
ISBN 3-540-43335-X Springer-Verlag Berlin Heidelberg New York

This work is subject to copyright. All rights are reserved, whether the whole or part
of the material is concerned, specifically the rights of translation, reprinting, re-use
of illustrations, recitation, broadcasting, reproduction on microfilms or in any other
way, and storage in data banks. Duplication of this publication or parts thereof is
permitted only under the provisions of the German Copyright Law of September 9,
1965, in its current version, and permission for use must always be obtained from
Springer-Verlag. Violations are liable for prosecution under the German Copyright
Law.

Springer-Verlag Berlin Heidelberg New York
a member of BertelsmannSpringer Science+Business Media GmbH

http://www.springer.de

© Springer-Verlag Berlin Heidelberg 2002
Printed in Germany

The use of general descriptive names, registered names, trademarks, etc. in this
publication does not imply, even in the absence of a specific statement, that such
names are exempt from the relevant protective laws and regulations and therefore
free for general use.

Typesetting: Camera ready by author
Cover design: *Erich Kirchner*, Heidelberg

Printed on acid-free paper   SPIN: 10869236   55/3142/du   5 4 3 2 1 0

Acknowledgements

Wolfgang Eichhorn deserves special thanks for his vital support as my Ph. D. supervisor. Siegfried Berninghaus, Jörg Gutsche, Armin Haas, Manfred Holler, Stefan Seifert, and Mika Widgrén were key motivators of my research and contributed to it in several ways. Material for this book also benefitted from suggestions made by Christian Bayer, Matthias Brueckner, Ingolf Dittmann, Karl-Martin Ehrhart, Susanne Fuchs-Seliger, Werner Güth, Burkhard Hehenkamp, Laurence Kranich, René Levínský, Moshé Machover, Atanasios Mitropoulos, Dilip Mookherjee, Alfred Müller, Anders Poulsen, Hannu Salonen, Dov Samet, Wendelin Schnedler, Alexander Tieman, and Karl-Heinz Waldmann. A number of seminar, summer school, workshop, and conference participants provided significant academic stimuli and Ingeborg Kast gave important organizational help. Thanks go to all of them. Moreover, I am particularly grateful to my family for their support – it was crucial for this work in many respects.

Karlsruhe, January 2002   *Stefan Napel*

Contents

Introduction

This book investigates bargaining between two agents. Its objective is to present, to extend, and to apply the present state of theoretical knowledge. A wide range of questions will be considered: First of all, will two parties reach efficient agreements? Traditional economic theory gives a generally affirmative answer for perfectly rational agents, who can carry out complex calculations instantaneously and without cost. The book uses innovative methods to analyse the implications of less demanding assumptions. A practical question related to bargaining is: How much power does the design of institutions such as the U. N. Security Council give to each of its members? Formally, non-permanent members' votes are necessary to pass resolutions, but theoretical investigation of pre-voting negotiation attributes all power to the five permanent members. Or one may ask whether a society should rather finance the education in higher mathematics for a talented person than remedial training for a retarded person? Different concepts of justice yield different answers. Which particular concept is implemented in a given society is also a matter of bargaining, and it is of special philosophical interest to investigate which bargain will be struck in an ideal society in which individual talents and resources are not yet known.

Very generally, a *bilateral bargaining situation* is characterized by two agents – individuals, firms, governments, etc. – who have a common interest in cooperating, but who have conflicting interests concerning the particular way of doing so. In economic terms, the agents can jointly produce some type of surplus, provided they agree on how to divide it. *Bilateral bargaining* refers to the corresponding attempt to resolve a bargaining situation, i. e. to determine the particular form of cooperation and the corresponding payoffs for both.

A lot of interaction in everyday's life is some type of bargaining. This includes quite trivial issues – for example agreeing when to meet, which restaurant to go to, etc. – as well as negotiation of industry-wide wages or details of European politics. In economics, a bargaining situation arises whenever two agents mutually enjoy some degree of monopoly power, i. e. if no perfect substitute for each other exists for economic, technological, legal, or other reasons. For example, this is the case if two firms or a firm and an employee have made relation-specific investments. It is also true if a seller and a buyer have a particular geographical or temporal preference for dealing with each other. Studies of bargaining hence complement and extend the theory of markets. Moreover, every non-bargaining interaction with player-specific best outcomes becomes a bargaining situation if it is repeated many times.

A *bargaining model* denotes some stylized representation of bargaining – mathematical, graphical, verbal, or other. Typically, bargaining involves strategic considerations. The chief analytical tool for strategic interaction is *game theory*, and the corresponding representations of bargaining are called *bargaining games*. They serve to distinguish ways of (non-)cooperation by the agents, also called *players*, as particularly plausible in a given context. Bargaining games give an explanation or reason for observed bargaining outcomes and allow to make predictions or even prescriptions.

Rather pragmatically, *bargaining theory* in this book refers to the collection of game-theoretic bargaining models. Its *application* will mean that a particular bargaining model is used in an encompassing model of economic or social behaviour – for example, wage setting or political decision making – in order to obtain new insights.

Chapter 1 presents the essentials of bargaining theory. It introduces the preliminaries for the extensions and applications in Chaps. 2–4. The presentation aims to combine mathematical precision with particular comprehensibility. The chapter is novel in describing the full *evolution of bargaining analysis* – starting with Francis Y. Edgeworth's (1881) first neoclassical treatment of individual and collective rationality in a bargaining situation, and ranging to contemporary investigations of boundedly rational agents. The chapter illuminates the links not only between traditional cooperative and non-cooperative game-theoretic analysis, but also the recent evolutionary approach to interaction. Moreover, its concluding section deals extensively with *experimental studies* of bargaining behaviour. Several concepts are presented and investigated in greater generality than is common. For example, sensitivity to a player's aversion to risk is proved for the asymmetric Nash solution and not the symmetric one; a compact formula is derived which describes the implication of a bargaining procedure with at most n stages of proposals and counter-proposals and which links the two standard bargaining models of a single ultimatum offer and of a potentially infinite sequence of offers.

Chapter 2 considers a low-rationality reference case for bilateral negotiations. An original model of satisficing, rather than optimizing bargainers is developed. It addresses *soft factors* of negotiation – emotional and irrational – which have so far not been studied in the economic literature. Novel links between agents' attributes such as persistence, stubbornness, or capriciousness and average bargaining success are established. Moreover, the implications of the satisficing heuristic for efficiency and distribution in repeated bargaining situations are deduced. Specifically, the model assumes both players to stick to their past action if it was satisfactory relative to a personal and adapting aspiration level. Otherwise, they possibly try a different strategy. Players' aspirations result from their individual average bargaining success, with more weight placed on the more recent past. They are also allowed to experience truly erratic jumps in their aspirations. The model can be used to explain empirical observations which question the rationalistic paradigm in game theory. Moreover, the model introduces new mathematical methods to the analysis of bargaining. Namely, the theory of Markov processes on general state spaces is applied in order to characterize long-run dynamics. It is proved that satisficing behaviour is 'rational' enough to yield (approximately) *efficient average surplus distributions*. Extensive computer simulations complement the mathematical investigations. They give insight into typical short-run bargaining dynamics, and yield numerical data about players' *parameter-dependent shares* of the available surplus.

Chapter 3 is devoted to an innovative application of bilateral bargaining theory to the measurement of economic and political decision power. Several power indices have been designed to capture the a priori distribution of power in decision bodies such as shareholder meetings or parliaments. However, it is pointed out that established indices overstate a player's power if some other player can credibly issue the following ultimatum to him: Accept (almost) no share of the spoils from a possible winning coalition or be prevented from taking part in one at all. This can deprive the former player of all influence on political or economic outcomes. A new classification of agents into *inferior players* and non-inferior players is developed in order to correct established indices. This allows for the axiomatic and probabilistic characterization of a new *strict power index*, which is better suited in many environments. This index is shown to have a desirable monotonicity property and to belong to a yet unexplored family of indices which allow flexible modelling of inferior players' decision behaviour. A new stability notion for coalitional games, which is based on a generalization of inferior players to inferior coalitions, is briefly explored. The chapter includes a concise introduction to main concerns, concepts, and conclusions of the power index literature. The notion of inferior players and the strict power index are based on recent work with Mika Widgrén.

The concluding Chap. 4 contributes an *analysis* of the more general credentials of established bargaining solutions as distribution norms. Bargaining situations are ubiquitous in social life. It is pointed out that many political institutions – such as compulsory social insurance, labour or rent legislation, etc. – and social norms – like offering a seat to an elderly person, splitting gains or costs fifty-fifty, or giving right of way to pedestrians – implement particular solutions to them. The different concepts of social justice and fairness which are implicitly captured by different bargaining solutions are identified. A common misjudgment concerning the prominent Nash bargaining solution is clarified. Namely, it reflects a balance of power, not any first moral principles concerning a 'fair arbitration scheme.' The appealing way to evaluate principles of social justice proposed by philosopher John Rawls (1971) is then briefly discussed. He considers the hypothetical negotiation of a social contract when positions in society are assumed to be afterwards randomly assigned to the bargainers. He does not use bargaining theory and makes the controversial decision-theoretic assumption that each player under these circumstances assumes to end up being the least advantaged member of society. Game theorist Ken Binmore's (1994, 1998b) recent attempt to reconstruct Rawls's conclusions in the framework of bargaining theory is therefore investigated critically. Advantages but also limits of approaching questions of social philosophy with theoretical models are evaluated. The chapter demonstrates that bargaining models – both of the traditional variant, and also unorthodox ones such as that developed in Chap. 2 – have implications reaching much farther than bargaining theorists' standard examples, like the division of €100, may suggest. Chapter 4 has benefitted from the collaboration with Manfred Holler.

The reader is assumed to be familiar with basic game theory. However, a brief summary of all necessary game-theoretic concepts and definitions is given in an appendix. Both for clearer exposition and more balanced usage of third-person singular pronouns, the text will refer to player 1 as female and to player 2 as male. For unspecified players, the pronouns 'he', 'his', etc. are used in a neutral sense.

1. Essentials of Bargaining Theory

This chapter gives an introduction to essential concepts and models of two-person negotiations, i. e. bilateral bargaining theory. The main purpose is to provide necessary preliminaries for the subsequent chapters. However, it is deemed worthwhile to add to an already large literature an introduction which aims to combine mathematical precision with particular comprehensibility, and which for the first time presents a comprehensive history of thought that ranges from Edgeworth's work in 1881 to most recent corroborations of classical predictions for bargaining by evolutionary models. Moreover, several results appear in more generality than is common or – to the author's knowledge – for the first time. Recommended other introductions to bargaining theory are Osborne and Rubinstein (1990), Binmore, Osborne, and Rubinstein (1992), and Muthoo (1999). Familiarity with basic game theory is assumed, but the appendix to this book collects all game-theoretic concepts, notation, and results which are used.

In the first section, Edgeworth's (1881) seminal formalization of individual and collective rationality in a bargaining situation is presented as well as Zeuthen's (1930) model of concession behaviour under risk and Hicks' (1932) equilibrium of strike resistances. Nash's (1950) axiomatic solution of the bargaining problem is the centerpiece of the subsequent section on cooperative models, which also covers links to the earlier contributions and alternative axiomatic solutions. A section on non-cooperative bargaining theory then deals primarily with the finite and infinite horizon alternating offers models of perfect information pioneered by Ståhl (1972) and Rubinstein (1982). The latter's relationship to Nash's axiomatic solution is pointed out. Some main features of models with incomplete information are illustrated briefly. In a section on evolutionary bargaining models it is then demonstrated that very demanding assumptions on players' rationality are not necessary for many predictions. The presentation covers the adaptive play model of Young (1993b), the imitation model of Gale, Binmore, and Samuelson (1995), and the model of finite bargaining automata proposed by Binmore, Piccione, and Samuelson (1998). The concluding section primarily deals with the link between theoretical predictions and observations of human bargaining behaviour in laboratory experiments.

1.1 Early Formalizations and Models

Bilateral exchange is a classical bargaining situation. Two agents, 1 and 2, have initial endowments $\bar{x}^1 = (\bar{x}^1_1, \bar{x}^1_2) \in \mathbb{R}^2_+$ and $\bar{x}^2 = (\bar{x}^2_1, \bar{x}^2_2) \in \mathbb{R}^2_+$ of two perfectly divisible goods. This defines a payoff combination $(\pi_1(\bar{x}^1), \pi_2(\bar{x}^2))$ which describes the *status quo* in terms of players' individual preferences, represented by utility functions π_i. Typically, this status quo can be improved by a (partial) bilateral exchange of goods. But what allocation will be reached if exchange is voluntary, the two agents are rational, and there is no exogenous arbitrator? Economists have been concerned with this question for a long time. A first formal answer is given by the Anglo-Irish economist Francis Y. Edgeworth in *Mathematical Psychics – An Essay on the Application of Mathematics to the Moral Sciences* (1881).

Edgeworth (implicitly) assumes that players' utility functions π_1 and π_2 are differentiable and sets out (p. 21)

> ... to find a point ... such that, in whatever direction we take an infinitely small step, ... [π_1 and π_2] do not increase together, but that, while one increases, the other decreases.

He defines the concept of a player's "line of indifference" (p. 21) – later a standard tool of economic analysis – and then observes that "... the direction which [player 1] ... will prefer to move ... is perpendicular to the line of indifference" (p. 22). He concludes that both players will prefer exchange as long as they can move in a direction "positive ... for both" (p. 22). Such a direction no longer exists if the players' indifference curves are tangential. Edgeworth derives

$$\frac{\partial \pi_1}{\partial x_1}\frac{\partial \pi_2}{\partial x_2} - \frac{\partial \pi_1}{\partial x_2}\frac{\partial \pi_2}{\partial x_1} = 0 \quad \Longleftrightarrow \quad \frac{\partial \pi_1/\partial x_2}{\partial \pi_1/\partial x_1} = \frac{\partial \pi_2/\partial x_2}{\partial \pi_2/\partial x_1} \qquad (1.1)$$

as the necessary (but not sufficient) tangency condition. This leads to his definition of the *contract curve* as the locus of all points x that satisfy (1.1).[1]

This situation can be illustrated with the *Pareto box*, a diagram often incorrectly attributed to Edgeworth, too.[2] Figure 1.1 superposes an (x^1_1, x^1_2)-diagram, in which indifference curves for player 1 are sketched, with a rotated (x^2_1, x^2_2)-diagram, which depicts indifference curves for player 2, in such a way that the edges of the resulting rectangle have length $(\bar{x}^1_1 + \bar{x}^2_1)$ and $(\bar{x}^1_2 + \bar{x}^2_2)$, respectively. The rectangle represents the *set of feasible allocations* of

---

[1] Some authors, like Mas-Colell, Whinston, and Green (1995, ch. 15), use a different terminology and equate the contract curve with the *bargaining set* or *core* defined below.

[2] The box is not depicted in Edgeworth (1881), and neither anywhere else in Edgeworth's work. Vilfredo Pareto uses it e.g. in his *Manuale di economia politica* in 1906 (cf. Pareto 1909, p. 191), several years before Bowley (1924, p. 5). The author thanks Wilhelm Lorenz for these references. Tarascio (1992) gives a most interesting account of the geneology of the Pareto box.

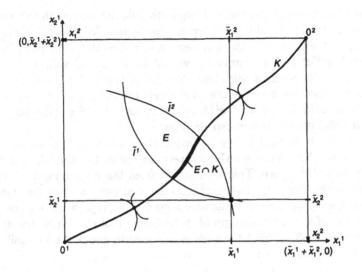

Fig. 1.1. Illustration of bilateral exchange in the Pareto box

goods based on players' initial endowments. Edgeworth's contract curve K is monotonically increasing from the origin of player 1, 0^1, to that of player 2, 0^2.

Points on the contract curve are Pareto-efficient. They can be characterized as *collectively rational*. Voluntary exchange requires that allocations are also *individually rational*, i. e. yield at least as much utility to either player as the respective initial endowment. In Fig. 1.1, this means that only allocations in the convex set E defined by player 1 and 2's indifference curves through the status quo, $\bar{I}^1$ and $\bar{I}^2$, should be considered.

Individual and collective rationality leave the set $E \cap K$ as possible equilibrium allocations. It is referred to as the *bargaining set* or the *core*.[3] This set is a small part of E, but typically not a singleton; it retains the multiplicity of mutually beneficial outcomes that characterizes a bargaining situation. Which particular allocation will be reached by voluntary, un-arbitrated bilateral exchange is therefore unspecified. In view of the remaining indeterminacy, Edgeworth (p. 30) quotes Jevons: "Such a transaction must be settled upon other than strictly economical grounds."[4]

---

[3] The bargaining set is distinct from the similarly named concept introduced by Aumann and Maschler (1964). The notions of collective rationality and individual rationality (with respect to all possible coalitions) also define the *core* of general n-player cooperative games.

[4] Edgeworth later on suggests that "the forces of self-interest being neutralised" on the contract curve, "competition requires to be supplemented by arbitration, and the basis of arbitration between self-interested contractors is the greatest possi-

John von Neumann and Oskar Morgenstern do not add much to this conclusion in their path-breaking *Theory of Games and Economic Behavior*. As an example, they consider the simplest case of *bilateral monopoly*, involving a single seller 1 of a commodity valued u by her and valued $v > u$ by a single buyer 2. Normalizing player 2's status quo utility to 0, von Neumann and Morgenstern derive the *characteristic function* $v: \wp(I) \to \mathbb{R}$ with $v(\{1\}) = u$, $v(\{2\}) = 0$, and $v(\{1,2\}) = v$, and thus formally describe a bargaining situation as a *non-zero-sum game*.

Economists have used utility representations of players' preferences long before, but von Neumann and Morgenstern are the first to consider the problem only in utility space. This abstraction from the underlying problem of trade, exchange, duopoly, wage setting, etc. produces the immense generality of the theory of games; it has also served as a preparation for the later first solution of the bargaining problem by Nash (1950a). Von Neumann and Morgenstern (1953, p. 555) themselves are skeptic about the possibility to reduce the bargaining set:

> There exists precisely one solution. It consists of all those imputations where each player gets individually at least that amount which he can secure for himself, while the two get together precisely the maximum amount which they can secure together.

In the example, there will be trade at a price p which satisfies

$$u \leq p \leq v, \tag{1.2}$$

and von Neumann and Morgenstern (p. 557) argue:

> Where p will actually be between the limits of [(1.2)] depends on factors not taken into account in this description . . . a satisfactory theory of this highly simplified model should leave the entire interval [(1.2)] available for p.

As has later been remarked by Harsanyi (1956, p. 145), the unpredictability claimed by von Neumann and Morgenstern is "by no means a law of nature, but is only a gap in current economic and political theories." In fact, there already existed two models of wage setting at the time of the *Theory of Games* which identify a unique outcome of bargaining.

The first one has been developed by Zeuthen (1930). It has been rediscovered and reappraised by Harsanyi (1956). Zeuthen argues in terms of two incompatible proposals $o_t^1 \neq o_t^2$ that have been made by player 1 and 2, respectively, at a given stage t of negotiations. Each player $i \in I = \{1, 2\}$ has as his most extreme choices to either accept the other player's proposal, o_t^{-i}, and to receive the payoff $\pi_i(o_t^{-i}) > 0$ for sure. Or player i can insist on his proposal o_t^i with the risk that, with some probability p_{-i}, player $-i$ leaves the

---

ble sum-total utility" (p. 56) – indicating support for the *utilitarian bargaining solution* (cf. Sect. 1.2.2 and Chap. 4).

table and pursues an outside option. Scaling utility[5] such that a breakdown of negotiations yields zero payoff, Zeuthen and Harsanyi argue that an expected utility maximizer i will accept proposal o_t^{-i} if $\pi_i(o_t^{-i}) > (1 - p_{-i})\pi_i(o_t^i)$.

This claim rests on questionable implicit assumptions, which must reduce the set of alternatives to either full or no concession. In any case, the utility quotient $\Delta\pi_i/\pi_i(t) := \left[\pi_i(o_t^i) - \pi_i(o_t^{-i})\right]/\pi_i(o_t^i)$ measures "the utmost probability of conflict to which [player i] can find its advantage to expose itself" (Zeuthen 1930, p. 115) by insisting on the better terms o_t^i instead of accepting the less favourable o_t^{-i}. This quotient is called player i's *risk limit* by Harsanyi. Intuitively, the risk limits "decide the strength of each party's 'determination'" (Harsanyi 1956, p. 148).

Two explicit assumptions are then made. First, player i will make a *concession* and propose an agreement o_{t+1}^i more favourable to $-i$ than o_t^i whenever

$$\frac{\Delta\pi_i}{\pi_i}(t) < \frac{\Delta\pi_{-i}}{\pi_{-i}}(t), \tag{1.3}$$

i.e. if he is less determined than $-i$. Player $-i$ sticks to his proposal if (1.3) is satisfied, so that $o_{t+1}^{-i} = o_t^{-i}$. Without loss of generality, i's concession can be considered big enough to reverse inequality (1.3). This leaves $-i$ the player who is more vulnerable to a threat of breakdown and who is to make a concession in $t + 1$. A possible motivation for this behaviour is given by Roth (1979, p. 30):

> Intuitively, think of the players sitting around the bargaining table after having made their most recent (incompatible) proposals. As time passes, the atmosphere grows more tense, and the probability of conflict rises, until one of the players (whose risk limit has been reached) breaks the deadlock by making a new proposal, with a more modest demand.

Second, if both players happen to have the same utility quotients, both will make some concession. Given a smallest monetary unit, or another technical or psychological lower bound to the size of a concession, the procedure will terminate after a finite number of steps.

It is doubtful whether Harsanyi's guess that (p. 151)

> Many economists ... will probably find Zeuthen's reasoning more convincing as it is based on a fairly plausible psychological model of the bargaining process, and ... look somewhat askance at Nash's game-theoretical method ...

is (still) accurate. Zeuthen's concession rule is not implausible, but a strong ad hoc assumption.[6] However, it will be seen in Sects. 1.2.1 and 1.3.2 that Zeuthen's behavioural model predicts approximately the same outcome as the game-theoretic investigations of Nash (1950a) and Rubinstein (1982).

---

[5] Zeuthen argues in terms of present values of money, not utility. The latter is introduced by Harsanyi.

[6] Harsanyi (1956) tries to derive Zeuthen's concession rule from more general behaviour postulates.

Another early model which singles out a unique bargaining outcome is proposed by Hicks (1932, ch. VII). Hicks argues that each player i is ready to accept a wage proposal, which is on the table, if it is less costly than a prospective labour dispute in case of rejection. One can construct a mapping from any (expected) length of a dispute to "the highest wage an employer will be willing to pay rather than endure a stoppage of that period" (p. 141). A higher proposed wage rate w implies a greater length l of labour dispute which is preferred to acceptance by the firm, and so the mapping – the *employer's concession curve* – is strictly increasing in l. Similarly, the union will associate each wage offer with a maximal length of strike such that it is indifferent between the two. Since a higher wage proposal by the employer implies a smaller benefit of a strike, the corresponding function – the *union's resistance curve* – is strictly decreasing in l. Under plausible assumptions the employer's concession curve and the union's resistance curve have a unique intersection. This defines a wage agreement w^* that makes each party just indifferent between accepting w^* or starting a labour dispute by a refusal.

Threats of strike or lockout are crucial features of wage bargaining. By requiring that such threats have to be credible, i.e. to reflect the true preference regarding wage settlement and stoppage of work, Hicks' verbal model has a flavour of the time-consistency requirement later formalized by subgame perfectness. Hicks supposes incomplete information in particular about the employer's concession curve. Because of the information asymmetry, he characterizes the uniquely defined w^* as "the highest wage which skilful negotiation can extract from the employer" (p. 144) – indicating that he does not feel to have determined a unique outcome of bargaining. Still, he deserves credit as one of the first economists to have solved the indeterminacy of bilateral monopoly.

1.2 Cooperative Solutions

Cooperative bargaining theory assumes that voluntary but binding agreements of any kind can be made by the players.[7] *Outcomes* $o \in O$ of bargaining are either particular cooperation agreements specifying a surplus distribution, or final disagreement. Any outcome determines a unique element of the *set of feasible payoff combinations* U. No specific bargaining rules are assumed. It is asked what sort of agreement tends to result for a given economic or political balance of power between two perfectly informed and rational agents. As argued by Edgeworth (1881), individual and collective rationality restrictions imply that candidate agreements lie in the bargaining set. If a bargaining situation is perfectly symmetric in utility terms, then a symmetric utility combination seems a natural prediction for the bargaining outcome. John

---

[7] A comprehensive overview of cooperative models of bargaining is given by Thomson (1994).

F. Nash (1950a) has been the first to provide a solution for a more general setting. Harsanyi (1956, p. 147) characterizes Nash's solution as "fundamentally a generalization of this [symmetry] principle for the more general case of asymmetric situations."

1.2.1 The Nash Bargaining Solution

Nash's breakthrough is achieved by the *axiomatic method*, which Nash (1953, p. 129) describes as follows:

> One states as axioms several properties that it would seem natural for the solution to have and then discovers that the axioms actually determine the solution uniquely.

It can be argued that often the method works exactly the other way round. In any case, Nash deduces a unique function that maps elements from a class of appropriately defined bargaining problems to a single element of the respective bargaining set by imposing certain reasonable requirements.

A *bargaining problem* in Nash's terminology is a pair $\langle U, u^D \rangle$ where $U \subset \mathbb{R}^2$ is the set of feasible payoff combinations and $u^D \in U$ is the *disagreement point* or *status quo point* which describes payoffs if no agreement is reached. U is taken to be convex, closed, and bounded from above,[8] and has at least one element $u \in U$ which is strictly preferred to u^D by both players, i.e. $u > u^D$.[9] All bargaining problems satisfying these requirements are collected in the set $\mathcal{B}$. Note that it is assumed to be of no concern to the players by what underlying bargaining situation the pair $\langle U, u^D \rangle$ is created, and that $\langle U, u^D \rangle$ is considered to be common knowledge to the players.

A *bargaining solution* can then be defined as a function $F \colon \mathcal{B} \to \mathbb{R}^2$ with $F(U, u^D) \in U$ which maps each bargaining problem to a unique feasible payoff vector. For given $\langle U, u^D \rangle$, $F(U, u^D)$ will be referred to as the *solution to bargaining problem* $\langle U, u^D \rangle$. Nash (1950a) identifies a particular function, F^N, called the *Nash (bargaining) solution*. In the following, the more general *asymmetric Nash (bargaining) solution* proposed by Kalai (1977a)[10] will directly be introduced and axiomatically characterized following the exposition in Binmore (1992, ch. 5.5).

---

[8] Nash assumes that U is compact. Also, he states more than four axioms; he lists eight in Nash (1950a), including assumptions which guarantee an expected utility representation of preferences, and seven in Nash (1953), including assumptions which are only relevant when u^D is not exogenously given.

[9] For $x, y \in \mathbb{R}^2$, let $x \geqq y$ denote $x_1 \geq y_1 \wedge x_2 \geq y_2$; let $x \geq y$ denote $x_i > y_i \wedge x_{-i} \geq y_{-i}$; and let $x > y$ denote $x_1 > y_1 \wedge x_2 > y_2$.

[10] One also finds the term *generalized Nash solution*. This, more accurately, refers to Harsanyi and Selten (1972), where an expression similar to the one obtained below is derived for the case of two bargainers with *incomplete preference information*. They, first, use a non-cooperative approach in the vein of Harsanyi (1967/68) to derive an "equilibrium set", from which, second, the unique solution point is selected – by axioms related to Nash's – which maximizes an n-factor Nash product with marginal type probabilities as exponents.

As argued above, it is desirable that a bargaining solution F selects an individually and collectively rational payoff combination. This is formalized by the following two axioms:

Individual rationality (IR):
For all $\langle U, u^D \rangle \in \mathcal{B}$

$$F(U, u^D) \geqq u^D.$$

Pareto efficiency (PAR):
For all $\langle U, u^D \rangle \in \mathcal{B}$

$$u \geq F(U, u^D) \implies u \notin U.$$

(IR) and (PAR) restrict the solution of any particular bargaining problem $\langle U, u^D \rangle$. Two more axioms are used in order to specify a certain consistency with respect to different, but related bargaining problems.

First, two bargaining problems $\langle U, u^D \rangle, \langle U', u^{D'} \rangle \in \mathcal{B}$ should be assigned the same outcome if they describe the same bargaining situation by different but equivalent von Neumann-Morgenstern utility representations of players' preferences. Equivalent preference representations differ by a *strictly increasing affine transformation* of payoffs, i. e. a mapping $\tau_i \colon \mathbb{R} \to \mathbb{R}$ with $\tau_i(u_i) = a u_i + b$ for constants $a, b \in \mathbb{R}$ and $a > 0$. Therefore, one can require:

Invariance to equivalent utility representations (INV):
Given $\tau(u) := (\tau_1(u_1), \tau_2(u_2))$ for any strictly increasing affine transformations τ_1 and τ_2, and any $\langle U, u^D \rangle \in \mathcal{B}$

$$F(\tau(U), \tau(u^D)) = \tau(F(U, u^D)).$$

Second, it can be argued that if $u^* = F(U, u^D)$ is the agreement reached by players given the problem $\langle U, u^D \rangle$ and if u^* is also a feasible payoff combination for the problem $\langle U', u^D \rangle$ with reduced payoff opportunities $U' \subset U$, the players should agree on u^* as the solution also to $\langle U', u^D \rangle$. This requirement supposes that all alternatives not chosen in a bargaining situation $\langle U, u^D \rangle$ are – except u^D – regarded as irrelevant by the players for finding the solution $F(U, u^D)$. The formal statement is:

Independence of irrelevant alternatives (IIA):
For all $\langle U, u^D \rangle \in \mathcal{B}$ and all $U' \subseteq U$ with $u^D \in U'$

$$F(U, u^D) \in U' \implies F(U', u^D) = F(U, u^D).$$

This axiom is, in particular, satisfied by any F that maximizes the value of some function on U or U', respectively.

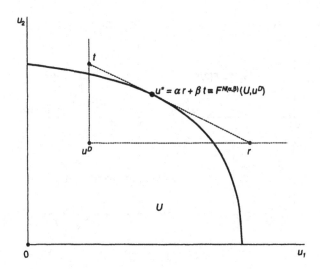

Fig. 1.2. The asymmetric Nash solution

Axioms (IR), (PAR), (INV), and (IIA) are *consistent*, i. e. do not reduce the set of suitable solutions F to the empty set. In particular, the *asymmetric Nash (bargaining) solution with bargaining powers α and β* (with $\alpha, \beta \geq 0$ and $\alpha + \beta = 1$) satisfies all four axioms. For given $\langle U, u^D \rangle$, it chooses the point u^* on the Pareto frontier of U such that the line $\overline{rt}$ through two points $r = (r_1, u_2^D)$ and $t = (u_1^D, t_2)$ is a supporting line to U at u^* where $u^* = \alpha r + \beta t$ (see Fig. 1.2). Let the function $F^{N(\alpha,\beta)}$ map each $\langle U, u^D \rangle \in \mathcal{B}$ to the point $u^* \in U$ uniquely defined in this way. Clearly, (IR), (PAR), and (IIA) are satisfied by construction. (INV) is satisfied since any affine transformation τ preserves convex structures, in particular the supporting-line property and the convex-combination property of $\overline{rt}$.

It turns out that not only does $F^{N(\alpha,\beta)}$ satisfy (IR), (PAR), (INV), and (IIA), but it is the only bargaining solution to do so, i. e. $F^{N(\alpha,\beta)}$ is *axiomatically characterized* by these properties:

Theorem 1.1. (Nash) *If $F\colon \mathcal{B} \to \mathbb{R}^2$ satisfies axioms (IR), (PAR), (INV), (IIA), then F is an asymmetric Nash bargaining solution for some non-negative bargaining powers α and β with $\alpha + \beta = 1$.*

Proof. Consider a bargaining solution F that satisfies (IR), (PAR), (INV), and (IIA). The canonic bargaining problem $\langle Z, 0 \rangle$ with $Z = \{u\colon u_1 + u_2 \leq 1\}$ and disagreement point $(0, 0)$ will first be investigated. By axioms (IR) and (PAR), F must specify a solution $u^{*\prime} = F(Z, 0)$ which can be expressed as

$$u^{*\prime} = \alpha r' + \beta t'$$

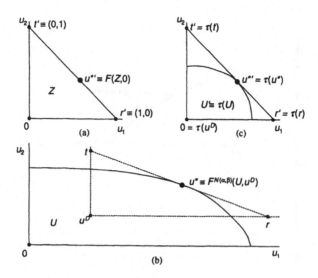

Fig. 1.3. Illustration of the proof of Theorem 1.1

for $r' = (1,0)$, $t' = (0,1)$, and a unique $(\alpha, \beta) \in [0,1]^2$ with $\alpha + \beta = 1$ (cf. Fig. 1.3 (a)).

For an arbitrary bargaining problem $\langle U, u^D \rangle \in \mathcal{B}$, let $u^* = F^{N(\alpha,\beta)}(U, u^D)$ be the asymmetric Nash bargaining solution with above α and β as bargaining powers. It uniquely defines points $r, t \in \mathbb{R}^2$ such that

$$u^* = \alpha r + \beta t$$

with $r_2 = u_2^D$ and $t_1 = u_1^D$ (see Fig. 1.3 (b)). Given these points, define the transformation $\tau \colon \mathbb{R}^2 \to \mathbb{R}^2$ with

$$\tau_1(u_1) = \frac{1}{r_1 - u_1^D} u_1 - \frac{u_1^D}{r_1 - u_1^D}$$

and

$$\tau_2(u_2) = \frac{1}{t_2 - u_2^D} u_2 - \frac{u_2^D}{t_2 - u_2^D}.$$

Clearly, $\tau(u^D) = 0$. Moreover, $\tau_1(r_1) = 1$ together with $\tau_2(r_2) = \tau_2(u_2^D) = 0$ implies $\tau(r) = r'$. Similarly, $\tau(t) = t'$. This implies $\tau(u^*) = \alpha\tau(r) + \beta\tau(t) = u^{*\prime}$, or

$$\tau\left(F^{N(\alpha,\beta)}(U, u^D) \right) = F(Z, 0). \tag{1.4}$$

Now consider the transformed set of feasible payoff combinations $U' = \tau(U)$. Since $\overline{rt}$ is a supporting line of U and gets transformed into the supporting line $\overline{r't'}$, which is the boundary of Z, the transformed set U' is a subset of Z containing $F(Z,0)$ (see Fig. 1.3 (c)). (IIA) then requires

$$F(Z,0) = F(U',0) = F\left(\tau(U), \tau(u^D)\right). \tag{1.5}$$

By (INV),
$$F\left(\tau(U), \tau(u^D)\right) = \tau\left(F(U, u^D)\right). \tag{1.6}$$

Equations (1.4)–(1.6) yield

$$\tau\left(F^{N(\alpha,\beta)}(U, u^D)\right) = \tau\left(F(U, u^D)\right). \tag{1.7}$$

The transformation τ is a bijective mapping. Therefore, (1.7) is equivalent to

$$F^{N(\alpha,\beta)}(U, u^D) = F(U, u^D).$$

Since $\langle U, u^D \rangle \in \mathcal{B}$ has been arbitrary, this proves the theorem. □

Nash (1950a) is only concerned with the symmetric case where α and β are equal ($\alpha = \beta = 1/2$). It is referred to as the *Nash (bargaining) solution* $F^N \equiv F^{N(\frac{1}{2},\frac{1}{2})}$. As demonstrated by Kalai (1977a), $F^{N(\alpha,\beta)}$ can (in the limit) be obtained from F^N by considering the n-player generalization of F^N (see p. 21) and replicating players 1 and 2 in a particular fashion. The symmetry underlying the Nash bargaining solution F^N is made precise by the following axiom:

Symmetry (SYM)
Let $\varrho(u_1, u_2) := (u_2, u_1)$. *Then for all* $(U, u^D) \in \mathcal{B}$

$$F\left(\varrho(U), \varrho(u^D)\right) = \varrho\left(F(U, u^D)\right).$$

As is clear from the proof of Theorem 1.1, the Nash bargaining solution is the unique bargaining solution which satisfies (SYM), (PAR), (INV), and (IIA).

The asymmetric Nash solution is conveniently characterized by the following result:

Theorem 1.2. (Nash) *For all* $\langle U, u^D \rangle \in \mathcal{B}$ *and all* $\alpha, \beta \geq 0$ *with* $\alpha + \beta = 1$ *the maximization problem*

$$\max_{u \in U, u \geq u^D} (u_1 - u_1^D)^\alpha (u_2 - u_2^D)^\beta \tag{1.8}$$

has a unique solution $u^* \in U$. *Moreover,*

$$u^* = F^{N(\alpha,\beta)}(U, d).$$

Proof. The set $\hat{U} = \{u \in U : u \geq u^D\}$ is compact. The objective function in (1.8) is continuous, and therefore has a global maximum on $\hat{U}$. Moreover, the function is strictly quasi-concave[11] and $\hat{U}$ is convex. Therefore, the maximizer u^* is unique.

---

[11] A real-valued function f is *strictly quasi-concave* on the open convex set $S \subseteq \mathbb{R}^n$ if for all $x \neq x' \in S$ and $\lambda \in (0,1) : f(x) \geq f(x') \Longrightarrow f(\lambda x + (1-\lambda)x') > f(x')$.

Next, consider the level curves of the objective function, implicitly defined by $g^c(u) = (u_1 - u_1^D)^\alpha (u_2 - u_2^D)^\beta - c = 0$ for $c \in \mathbb{R}_+$. These cannot intersect, i. e. each $u \in U$ belongs to exactly one level curve. For arbitrary $\hat{u} \in \mathbb{R}^2$ with $g^c(\hat{u}) = 0$, the tangent line to the graph of g^c at $\hat{u}$ is implicitly given by $\nabla g^c(\hat{u}) \cdot (u - \hat{u}) = 0$ (see e. g. Sydsæter, Strøm, and Berck 1999, p. 30). In particular, the tangent line at u^* is defined by

$$\alpha \left(\frac{u_1 - u_1^*}{u_1^* - u_1^D} \right) + \beta \left(\frac{u_2 - u_2^*}{u_2^* - u_2^D} \right) = 0. \tag{1.9}$$

Define r and t to be the intersections of this tangent line and the lines given by $u_2 = u_2^D$ and $u_1 = u_1^D$, respectively. Then, (1.9) is equivalent to

$$\alpha \left(\frac{u_1 - u_1^*}{u_1^* - t_1} \right) + \beta \left(\frac{u_2 - u_2^*}{u_2^* - r_2} \right) = 0. \tag{1.10}$$

Since r and t by above choice must both satisfy (1.10), one obtains

$$\alpha \left(\frac{r_1 - u_1^*}{u_1^* - t_1} \right) - \beta = 0 \quad \text{and} \quad -\alpha + \beta \left(\frac{t_2 - u_2^*}{u_2^* - r_2} \right) = 0$$

or

$$u_1^* = \alpha r_1 + \beta t_1 \quad \text{and} \quad u_2^* = \alpha r_2 + \beta t_2.$$

Thus, with $\alpha + \beta = 1$, $u^* = F^{N(\alpha, \beta)}(U, u^D)$. □

The objective function in Theorem 1.2 is called the *(generalized) Nash product*. Given $a, b \geq 0$ with $a + b > 0$ and $u^D \in U$, the expression $(u_1 - u_1^D)^a (u_2 - u_2^D)^b$ has the same maximizer u^* as the Nash product $(u_1 - u_1^D)^\alpha (u_2 - u_2^D)^\beta$ for $\alpha = a/(a + b)$ and $\beta = b/(a + b)$. By Theorem 1.2, the *asymmetric Nash solution with bargaining powers α and β* can therefore be defined for the weaker restriction of bargaining powers $\alpha, \beta \geq 0$ and $\alpha + \beta > 0$ by

$$F^{N(\alpha, \beta)}(U, u^D) := \underset{u \in U, u \geq u^D}{\arg\max} \, (u_1 - u_1^D)^\alpha (u_2 - u_2^D)^\beta$$

for $\langle U, u^D \rangle \in \mathcal{B}$. Similarly, the *Nash solution* has the explicit functional form

$$F^N(U, u^D) := \underset{u \in U, u \geq u^D}{\arg\max} \, (u_1 - u_1^D)(u_2 - u_2^D).$$

Based on Theorem 1.2, a useful characterization of the asymmetric Nash solution exists when the Pareto frontier $P(U)$ is the graph of a differentiable function $\phi \colon [\underline{u}_1, \overline{u}_1] \to [\underline{u}_2, \overline{u}_2]$, i. e. $P(U) = \{(u_1, \phi(u_1)) : u_1 \in [\underline{u}_1, \overline{u}_1]\}$, where $\underline{u}_i = \min\{u_i : u \in P(U)\}$ and $\overline{u}_i = \max\{u_i : u \in P(U)\}$ (cf. e. g. Muthoo 1999, pp. 24 and 36).[12]

---

[12] ϕ is strictly decreasing by the definition of $P(U)$, and concave because U has been assumed to be convex.

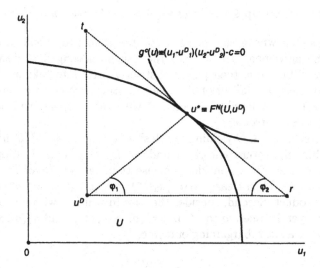

Fig. 1.4. Tangency criterion for the symmetric Nash solution

Lemma 1.1. (Tangency criterion) *For any $\langle U, u^D \rangle \in \mathcal{B}$ with $u_i^D \geq \underline{u}_i$ for $i = 1, 2$, the asymmetric Nash solution with bargaining powers α and β $(\alpha, \beta > 0)$ of bargaining problem $\langle U, u^D \rangle$ is the unique solution u^* to the equations*

$$-\phi'(u_1^*) = \frac{\alpha\,(u_2^* - u_2^D)}{\beta\,(u_1^* - u_1^D)} \tag{1.11}$$

and $u_2^* = \phi(u_1^*)$, where ϕ' denotes the first derivative of ϕ.

This is easily seen by considering the first-order condition of the maximization problem in Theorem 1.2. Lemma 1.1 has an appealing graphical implication for the symmetric case $\alpha = \beta$. The right-hand side of (1.11) denotes the ratio of players' net gains and defines the slope of line $\overline{u^D u^*}$. The left-hand side is the negative of the slope of the tangent line to U at u^*.[13] Thus, $\overline{rt}$ is the tangent line to U such that angles φ_1 and φ_2 in Fig. 1.4 are identical.[14]

The Nash solution has the following interpretation in terms of outcomes $o \in O$ underlying the payoff combinations in U (cf. Osborne and Rubinstein 1990, pp. 16ff). For convenience, choose both players' preference representations such that the utility of disagreement is zero. Now, consider a bargaining outcome $o^* \in O$ with the following stability property:

---

[13] It can be interpreted as the marginal rate of utility transfer between players in u^*, which is a constant in case of linearly transferable utility.

[14] When ϕ is not differentiable the Nash solution is nevertheless characterized as the solution which selects the unique u^* such that $\overline{u^D u^*}$ has the negative slope of some supporting line of U at u^* (cf. Roth 1979, pp. 43ff).

$(\forall i \in I) \, (\forall o \in O) \, (\forall p \in [0,1]) \colon \{ p \, \pi_i(o) > \pi_i(o^*) \implies p \, \pi_{-i}(o^*) > \pi_{-i}(o) \}.$

This means that whenever a player i strictly prefers to object to o^* and to propose the agreement $o \in O$ with the risk of causing final disagreement with probability $1 - p$, then player $-i$ strictly prefers to take an analogous risk and to reject o in favour of o^* while accepting that negotiations fail with probability $1 - p$. A feasible outcome o^* with this property will be called a *Nash bargaining outcome*.

Any agreement $o \in O$ yielding the Nash solution vector $F^N(U, u^D) = \pi(o)$ is a Nash bargaining outcome of the underlying bargaining situation. The reverse is also true: First, consider the case that neither player $i \in I$ prefers any $o \in O$ to o^*. Then clearly $\pi_1(o^*)\pi_2(o^*) \geq \pi_1(o)\pi_2(o)$, and o^* maximizes the Nash product. Second, consider the case in which – without loss of generality – player 1 prefers o to o^*, i.e. $\pi_1(o) > \pi_1(o^*)$, and $\pi_1(o), \pi_2(o) > 0$. Because o^* is a Nash bargaining outcome,

$$ p > \frac{\pi_2(o)}{\pi_2(o^*)} $$

holds for all $p > \frac{\pi_1(o^*)}{\pi_1(o)}$.[15] This implies

$$ \frac{\pi_1(o^*)}{\pi_1(o)} \geq \frac{\pi_2(o)}{\pi_2(o^*)}, $$

showing that o^* has a maximal Nash product.

A very close relation between the Nash solution and the Zeuthen-Harsanyi procedure of Sect. 1.1 exists. In fact, (1.3) on p. 9 is equivalent to

$$ \pi_i(o_t^i)\pi_{-i}(o_t^i) < \pi_i(o_t^{-i})\pi_{-i}(o^{-i}), $$

i.e. the player i whose last offer has the lower Nash product[16] makes a concession. He puts forward a new proposal o_{t+1}^i with a higher Nash product than o_t^{-i} to reverse the inequality. The process terminates precisely when the Nash product is maximized (subject to the indivisibility constraint on the concessions). This means that the behavioural approach of Zeuthen and the axiomatic approach by Nash predict essentially the same bargaining outcome – a fact first pointed out by Harsanyi (1956).

Players' attitudes towards risk are a crucial determinant of the Nash solution. Consider two von Neumann-Morgenstern utility functions π_i and $\tilde{\pi}_i$ which represent exactly the same preferences with respect to deterministic outcomes but not lotteries of outcomes. Then, $\tilde{\pi}_i$ is (strictly) *more risk averse* than π_i if for all deterministic outcomes $o \in O$, $\tilde{\pi}_i(o) = k(\pi_i(o))$ for a strictly

---

[15] Certainly, such a p exists if $\pi_1(o) > \pi_1(o^*)$.

[16] Recall that utility functions π_i have been chosen such that $\pi(u^D) = (0,0)$.

increasing (strictly) concave function k.[17] Now imagine that in a given bargaining situation player 2 with utility function π_2 is replaced by a player $2'$ who is more risk averse, i. e. has the utility function $k(\pi_2)$ for increasing concave k. It can be checked that the risk limit of player $2'$ is smaller than that of player 2 for any two deterministic proposals $o^i, o^{-i} \in O$ which are on the table. Thus player $2'$ always makes larger concessions to player 1 than player 2 does. This suggests that the more risk-averse her opponent is, the greater the utility which player 1 gets from bargaining over deterministic outcomes is. For the symmetric Nash solution Kihlstrom, Roth, and Schmeidler (1981) have derived this *risk sensitivity* of the Nash bargaining solution when the Pareto frontier of U is the graph of a differentiable function ϕ as in Lemma 1.1. The asymmetric solution is similarly risk-sensitive:

Proposition 1.1. *Consider a convex, compact set of deterministic outcomes $O \subset \mathbb{R}^n$ with associated lotteries $\Delta(O)$ and the bargaining problems $\langle U, u^D \rangle, \langle U', u^{D\prime} \rangle \in B$ defined by $\Delta(O)$, the disagreement outcome $o^D \in O$, player 1's weakly concave expected utility function π_1 and player 2's weakly concave expected utility function π_2 and π_2', respectively. If $\pi_2' = k(\pi_2)$ for a strictly increasing, concave, and differentiable function k, then*

$$F_1^{N(\alpha,\beta)}(U', u^{D\prime}) \geq F_1^{N(\alpha,\beta)}(U, u^D)$$

for $\alpha, \beta > 0$, i. e. player 1 benefits from bargaining with a more risk-averse player 2.

Proof. Without loss of generality, choose $u^D = u^{D\prime} = \pi(o^D) = \pi'(o^D) = (0,0)$. The asymmetric Nash solution of $\langle U', u^{D\prime} \rangle$ maximizes

$$f(u_1) = u_1^\alpha \, k \, (\phi(u_1))^\beta$$

for increasing concave k. Let u_1^* be player 1's utility from the Nash solution of $\langle U, u^D \rangle$. At $u_1 = u_1^*$, f has the first derivative

$$f'(u_1^*) = \alpha \, u_1^{*\alpha-1} \, k \, (\phi(u_1^*))^\beta + u_1^{*\alpha-1} \, u_1^* \, \beta \, \phi'(u_1^*) \, k' \, (\phi(u_1^*)) \, k \, (\phi(u_1^*))^{\beta-1} .$$
$$(1.12)$$

By Lemma 1.1, $u_1^* \beta \phi'(u_1^*) = -\alpha \phi(u_1^*)$. Hence, (1.12) becomes

$$f'(u_1^*) = \alpha \, u_1^{*\alpha-1} \, k \, (\phi(u_1^*))^{\beta-1} \, [k \, (\phi(u_1^*)) - \phi(u_1^*) \, k' \, (\phi(u_1^*))]$$
$$= \alpha \, u_1^{*\alpha-1} \, k \, (\phi(u_1^*))^{\beta-1} \, \phi(u_1^*) \left[\frac{k(\phi(u_1^*))}{\phi(u_1^*)} - k' \, (\phi(u_1^*)) \right] .$$

By the concavity of k, the last factor is non-negative. The product in front is positive because the Nash solution satisfies (IR) strictly for $\alpha, \beta > 0$. Hence, an increase of player 1's utility u_1^* weakly increases the Nash product for the problem $\langle U', u^{D\prime} \rangle$, implying that $F_1^{N(\alpha,\beta)}(U', u^{D\prime}) \geq F_1^{N(\alpha,\beta)}(U, u^D)$. $\square$

---

[17] Cf. e. g. Roth 1979, pp. 35ff for details. Not all utility functions are comparable in this fashion.

This result has an alternative interpretation. An increasing concave transformation to π_2 also results when player 2's marginal utility for the underlying
commodity or money is made to decrease more rapidly. Consider, for example,
the division of €100. Two individuals with identical strictly concave utility of
money and same initial wealth will divide the amount equally. But if player 1
has smaller decreases in marginal utility – or even has a constant marginal
utility – she will receive more than half. If both have identical preferences
but player 1 has higher initial wealth, she experiences a smaller decline of
marginal utility with respect to every cent of the €100, too – resulting in a
larger share for the richer player 1.

The intuition for Proposition 1.1 is that fear of disagreement causes a
highly risk-averse player to settle for more unfavourable (deterministic) agreements than the same player would if he were less risk averse. This intuition
and also Proposition 1.1 do not generally hold in case that agreements are
risky prospects.[18] For illustration, assume that $O = \{o^1, o^2, o^D\}$, where o^i
is player i's most preferred outcome. Then the Pareto frontiers $P(U)$ and
$P(U')$ are necessarily linear because expected utility is linear in probabilities (cf. Fig. 1.5), and involve non-trivial lotteries for all payoff combinations
except $\pi(o^1)$ and $\pi(o^2)$. Increased risk-aversion of player 2 can no longer be
studied by making $P(U)$ more concave, but rather u_2^D has to be shifted.
Suppose that player 2 strictly prefers o^D to o^1, i.e. $u_2^D > \pi_2(o^1)$. If he becomes strictly more risk averse, then a lottery between o^1 and o^2 must place
greater probability on o^2 in order to be preferred to the deterministic disagreement outcome o^D. If utilities are scaled such that $\pi_2(o^1) = \pi_2'(o^1) = 0$
and $\pi_2(o^2) = \pi_2'(o^2) = 1$, this implies $u^{D'} = \pi_2'(o^D) > u^D = \pi_2(o^D)$. The
Nash solution will then give *smaller* utility to player 1 in problem $\langle U', u^{D'} \rangle$
than in problem $\langle U, u^D \rangle$. So in contrast to Proposition 1.1, player 1 loses
from bargaining with a more risk-averse player 2. The implication of Proposition 1.1 only holds, in this simple example, if player 2 prefers o^1 to o^D
(corresponding to status quo $\bar{u}^D$ in Fig. 1.5).

Roth and Rothblum (1982) more generally investigate the Nash solution
when the set of deterministic outcomes, O, is not convex, so that some Pareto-
efficient expected utility combinations can only be obtained by lotteries. As
above, it may be detrimental to face a more risk-averse player in this setting.
Roth and Rothblum derive sufficient conditions for the bargaining problem,
underlying deterministic outcomes, and preferences such that the Nash solution either is traditionally risk sensitive as in Proposition 1.1, or exhibits
reverse risk sensitivity.

So far, players' disagreement payoffs, u^D, have been considered to be exogenously given. Nash (1953) investigates a model of *bargaining with variable
threats*. There, players choose (possibly mixed) *threat strategies* σ_i before

---

[18] The assumption of a convex, compact set of deterministic outcomes $O \subset \mathbf{R}^n$ and
concave utility functions ensures that each payoff combination in U or U' can be
obtained without resorting to non-degenerate lotteries $l \in \Delta(O)$.

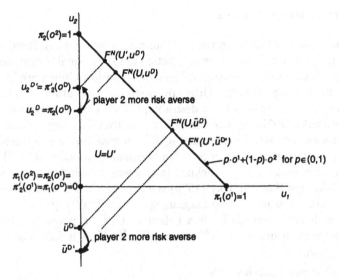

Fig. 1.5. The Nash solution and risk aversion for deterministic outcomes o^1, o^2, and o^D

the actual negotiations begin. If negotiations should fail, these threats must be carried out – whether they constitute an equilibrium or not – and jointly determine $u^D(\sigma)$. This two-stage game has a Nash equilibrium, which is characterized by a profile σ^* of *optimal threats* and the subsequent agreement on the (fixed-threat) Nash solution given $u^D(\sigma^*)$. A more detailed exposition of Nash bargaining with variable threats is, for example, given by Holler (1992, ch. 5).

Above presentation is concerned only with the case of bilateral bargaining. It is noteworthy that essentially all results can be extended to the case where all of n players have to agree on one of many mutually beneficial outcomes. The corresponding *n-player asymmetric Nash solution* is simply

$$F^{N(\alpha_1,\ldots,\alpha_n)} := \arg\max_{u\in U,\, u\geqq u^D} \prod_{i=1}^{n}(u_i - u_i^D)^{\alpha_i}$$

for a convex closed set of feasible payoffs $U \subset \mathbb{R}^n$ bounded from above, disagreement point $u^D \in U$, and bargaining powers $\alpha_1,\ldots,\alpha_n \geq 0$ ($\sum a_i > 0$). Note that this solution is based on the assumption that proper sub-coalitions of $I = \{1,\ldots,n\}$ cannot make mutually beneficial partial agreements. The n-player case in which arbitrary coalitions $S \subseteq I$ can form to mutual benefit is more complex. In this context, a prominent solution concept is the *Shapley value* (Shapley 1953). When partial agreements are not mutually beneficial, the disagreement payoffs are given, and utility is linearly transferable, it coincides with the n-player Nash solution.

1.2.2 Alternative Solutions

If one agrees that the axioms in the previous section adequately describe what
outcome can be expected from the interaction of two rational bargainers, Theorem 1.1 provides a definite answer on how this translates into a prediction for
any given bargaining problem. Other properties of a solution can, however,
be argued to be more plausible (or desirable) than Nash's. For example, Kalai
and Smorodinsky (1975) replace the rather controversial independence axiom
(IIA) with a monotonicity axiom. Let $u_i^B(U) := \max \{u_i : u \in U\}$ denote the
best that player i can possibly hope for in bargaining situation $\langle U, u^D \rangle \in \mathcal{B}$.[19]
The payoff combination $u^B(U)$ defined in this way is called the *ideal point* of
$\langle U, u^D \rangle$. Kalai and Smorodinsky (1975) require that if the ideal payoff $u_{-i}^B(\cdot)$
of player $-i$ coincides in the two bargaining games $\langle U, u^D \rangle, \langle U', u^D \rangle \in \mathcal{B}$ and
if the set of feasible payoffs U' is a subset of U, then player i will receive
at least as much utility in $\langle U, u^D \rangle$ as in $\langle U', u^D \rangle$. This is formalized by the
following axiom:

Individual monotonicity (IM):
For all $\langle U, u^D \rangle, \langle U', u^D \rangle \in \mathcal{B}$

$$u_{-i}^B(U) = u_{-i}^B(U') \ \wedge \ U' \subseteq U \implies F_i(U, u^D) \geq F_i(U', u^D).$$

Kalai and Smorodinsky (1975) show that the unique solution which satisfies (PAR), (SYM), (INV), and (IM) is the function F^{KS} – also called the
Kalai-Smorodinsky (bargaining) solution – which maps $\langle U, u^D \rangle \in \mathcal{B}$ to the intersection of the line $\overline{u^D u^B(U)}$ and U's Pareto frontier $P(U)$ (see Fig. 1.6).[20]
Because Nash and Kalai-Smorodinsky solutions both satisfy (PAR), (SYM)
and (INV), they coincide for symmetric games and games which can be transformed into a symmetric game by applying an increasing affine function. The
Kalai-Smorodinsky solution exhibits the same risk sensitivity as the Nash
solution (cf. Kihlstrom, Roth, and Schmeidler 1981).

F^{KS} has some credentials as identifying a 'fair' bargaining outcome (cf.
Chap. 4). Independent of each player's von Neumann-Morgenstern preference representation, it assigns to each player exactly the same percentage
achievement of his maximal possible gain on the status quo. More formally,

$$\frac{u_1^* - u_1^D}{u_1^B(U) - u_1^D} = \frac{u_2^* - u_2^D}{u_2^B(U) - u_2^D}$$

---

[19] An alternative definition of u_i^B is used by Roth (1979, p. 99) and many authors
thereafter, namely $u_i^B(U, d) = \max \{u_i : u \in U \wedge u \geq u^D\}$. In their original paper, Kalai and Smorodinsky do not require u_i^B to be obtainable by some $u \in U$
which is individually rational for $-i$.

[20] F^{KS} can straightforwardly be extended to $n \geq 3$ players. However, without
modification, it violates (PAR) in this case.

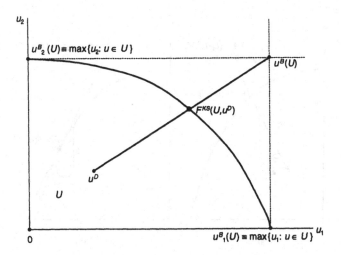

Fig. 1.6. The Kalai-Smorodinsky bargaining solution F^{KS}

holds for $u^* = F^{KS}(U, d)$, where both ratios are unaffected by arbitrary strictly increasing affine transformations τ_1 and τ_2 of u_1 and u_2, respectively.

Despite satisfying (IM), the Kalai-Smordinsky solution is not monotonic in the following, quite natural sense:

Monotonicity (MON):
For all $\langle U, u^D \rangle, \langle U', u^D \rangle \in \mathcal{B}$

$$ U' \subseteq U \implies F(U, u^D) \geqq F(U', u^D). $$

This monotonicity axiom is satisfied by the *weighted egalitarian* or *proportional (bargaining) solution* $F^{E(a,b)}$ with weights $a, b \geq 0$ and $a + b > 0$. It maps each bargaining problem $\langle U, u^D \rangle \in \mathcal{B}$ to the maximal element $u^* \in U$ of the line through u^D with slope b/a (cf. Fig. 1.7 (a)). In case of symmetric weights $a = b$, it is referred to as the *egalitarian (bargaining) solution* F^E. Assuming that one incremental unit of player 1's utility is comparable to one incremental unit of player 2's utility, F^E gives both players an equal gain relative to the status quo u^D. The same applies to $F^{E(a,b)}$ when a incremental units of player 1's utility correspond to b incremental units of player 2's utility. (MON) is implied by the following stronger property:[21]

Decomposability (DEC):
For all $\langle U, u^D \rangle, \langle U', u^D \rangle \in \mathcal{B}$

---

[21] To see that (DEC) is in fact stronger, consider the monotonic solution defined by mapping $\langle U, u^D \rangle$ to the maximal element $u^* \in U$ of the increasing and strictly concave curve through u^D with $u_2(u_1) = u_2^D + \ln(u_1 - u_1^D + 1)$.

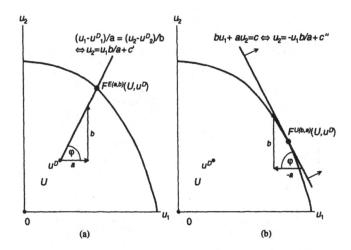

Fig. 1.7. Weighted egalitarian and utilitarian solutions

$$U' \subseteq U \implies F(U, u^D) = F\left(U, F(U', u^D)\right).$$

Axiom (DEC) states that bargaining can be decomposed into an arbitrary number of steps – each with a larger set of feasible payoffs and the new disagreement point being the status quo reached in the last step – without changing the final utility allocation. $F^{E(a,b)}$ satisfies (DEC). It is essentially the only bargaining solution to do so. However, $F^{E(a,b)}$ violates (PAR).[22] This illustrates the common trade-off between equality of outcome and efficiency.

The *weighted utilitarian (bargaining) solution* $F^{U(b,a)}$ with weights $b, a \geq 0$ and $b + a > 0$ maps each bargaining problem $\langle U, u^D \rangle \in \mathcal{B}$ to the unique maximizer of $bu_1 + au_2$ on $\hat{U} = \{u \in U : u \geq u^D\}$ which is closest to the line through u^D with slope b/a.[23] The simple case of $b = a = 1$ is referred to as the *utilitarian (bargaining) solution* F^U. Assuming that one incremental unit of player 1's utility is comparable to one incremental unit of player 2's utility, F^U maximizes the total utility gain of both players relative to the status quo u^D and also the total utility of both players subject to the individual rationality constraint. The same applies to $F^{U(b,a)}$ when a incremental units of player 1's utility correspond to b incremental units of player 2's utility because

---

[22] If U is u^D-comprehensive, i. e. $u \in U \wedge u^D \leq u' \leq u \implies u' \in U$, then it satisfies *weak Pareto efficiency*, i. e. $u > F^{E(a,b)}(U, u^D) \implies u \notin U$.

[23] The latter applies only if the Pareto frontier of U happens to have a linear part with slope $-b/a$. Equality of re-scaled utility is then called for as a reasonable way to break ties.

$$\arg\max_{(u_1,u_2)\in\hat{U}} \left(\frac{u_1 - u_1^D}{a} + \frac{u_2 - u_2^D}{b}\right) = \arg\max_{(u_1,u_2)\in\hat{U}} \left(bu_1 + au_2\right).$$

Figure 1.7 (b) illustrates the utilitarian solution $F^{U(b,a)}$. It is obvious from the angles indicated in figures 1.4 and 1.7 that if two of the solutions $F^N(U,u^D)$, $F^{E(a,b)}(U,u^D)$, and $F^{U(b,a)}(U,u^D)$ coincide for some $\langle U,u^D\rangle$, then the third is located at the same point as the other two.

Returning to the Nash bargaining solution F^N, note that the four axioms (PAR), (SYM), (INV), and (IIA) used for its characterization are *logically independent* and none is superfluous. This is proved as follows: F^{KS} satisfies (PAR), (SYM), and (INV) – but not (IIA). F^U satisfies (PAR), (SYM), and (IIA) – but not (INV). $F^{N(a,b)}$ for $a \neq b$ satisfies (PAR), (INV), and (IIA) – but not (SYM). Finally, the trivial solution with $F \equiv u^D$ satisfies (SYM), (INV), and (IIA) – but clearly not (PAR).[24]

1.3 Non-cooperative Models

Cooperative solutions attempt a prediction of what binding agreement two bargainers can generally be expected to reach in an unspecified negotiation process. *Non-cooperative models of bargaining*, in contrast, consider bargaining as a fully specified game; any agreements prior to playing the game are assumed to be non-binding and are hence ignored. Nash (1953, p. 129) describes the relation to the cooperative approach to bargaining as follows:

> ... one makes the player's steps of negotiation in the cooperative game become moves in the non-cooperative model. Of course, one cannot represent all possible bargaining devices as moves in the non-cooperative game. The negotiation process must be formalized and restricted, but in such a way that each participant is still able to utilize all the essential strengths of his position.

He continues:[25]

> The two approaches to the problem, via the negotiation model or via the axioms, are complementary; each helps to justify and clarify the other.

Cooperative solutions' level of abstraction is both strength in terms of rather general predictions, and weakness by offering few criteria of when these predictions apply. Similarly, non-cooperative models' complete specification, on the one hand, is a highly desirable feature. On the other hand, predictions – made in terms of Nash equilibria and its refinements – are sometimes extremely sensitive to even slight changes in the game form.

---

[24] If this trivial solution is excluded, F^N could also be characterized by replacing (PAR) with (IR), and $F^{N(\alpha,\beta)}$ by just requiring (IR), (INV), and (IIA).

[25] The related research agenda which attempts to provide explicit links between axiomatic cooperative solutions and fully specified non-cooperative models and their equilibria has been named the *Nash programme* (cf. Binmore and Dasgupta 1987b).

1.3.1 Games with Finite Horizon

One of the simplest non-cooperative models of bilateral bargaining is the *ultimatum game* illustrated in Fig. 1.8 (a). Players 1 and 2 can share a surplus of one unit provided they agree on a particular division of it. Player 1 is the first to move by making a proposal $x \in [0,1]$ which denotes the surplus division $(x, 1-x)$; she is referred to as the *proposer*. Player 2 responds to proposal x by either rejecting or accepting it; he is called the *responder*. For simplicity, let players' utility functions be linear in x and $1-x$, respectively.[26] Only pure strategies will be considered.

The reduced normal form of the ultimatum game is given by the set $I = \{1,2\}$ of players, the proposer's strategy space $S_1 = [0,1]$, the responder's strategy space $S_2 = \{s \mid s \colon S_1 \to \{0,1\}\}$ where 0 denotes rejection and 1 acceptance of player 1's proposal, and the payoff functions $\pi_1(s_1, s_2) = s_1 \, s_2(s_1)$ and $\pi_2(s_1, s_2) = (1-s_1)s_2(s_1)$. Consider the family of response strategies with

$$s_2^a(s_1) = \begin{cases} 0; \ 1 - s_1 < a \\ 1; \ 1 - s_1 \ge a \end{cases}$$

for $a \in [0,1]$, which formalize that player 2 wants at least the amount a. Every profile $(s_1, s_2^{1-s_1}) \in S_1 \times S_2$ constitutes a Nash equilibrium (NE) of the ultimatum game. Thus any efficient surplus division is supported by a NE.

A clearer prediction is obtained by using a refined equilibrium notion. Namely, the ultimatum game has a unique subgame perfect equilibrium (SPE). To see this, note that any strategy $s_2 \in S_2$ which prescribes rejection in case of a proposal $s_1 \in [0,1)$ does not induce a Nash equilibrium in the subgame that starts after s_1 has been proposed. Player 2's strategy would constitute an *incredible threat* if it specified that an offer which gives him a strictly positive amount $1 - s_1$ is rejected; the SPE concept rules out such incredible threats (and incredible promises). The only candidates for player 2's strategy in a SPE are therefore s_2^0, and the strategy s_2' which rejects $s_1 = 1$ but is otherwise identical to s_2^0. The profile $s^* = (1, s_2^0)$ is a SPE – specifying that player 1 asks for the entire surplus, and player 2 accepts.[27] The profile $(1, s_2')$ is no SPE since it is sub-optimal for player 1 to propose $s_1 = 1$, which is rejected by s_2', anticipating that any proposal $s_1 < 1$ would be accepted by 2. Since, however, no such proposal s_1 is optimal – max $[0,1)$ does not exist – s^* is the unique SPE of the ultimatum game.[28]

---

[26] The results for the ultimatum game do not change if players have arbitrary preferences as long as a greater share is more desirable than a smaller one.

[27] Acceptance by player 2 is only a weak best response. This Nash equilibrium is therefore not strict.

[28] If proposals have to be made in multiples of a smallest – e. g. monetary – unit ξ, then $s^{**} = (1 - \xi, s_2^\xi)$ is the only other SPE. So, the game-theoretic prediction of an extremely asymmetric division is robust in this case.

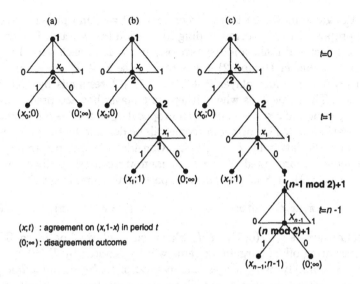

Fig. 1.8. Ultimatum game form, 2-stage and n-stage alternating offers bargaining game forms

This quite extreme SPE prediction – usually interpreted as the game-theoretic prediction per se – indicates overwhelming *bargaining power* of the proposer in the ultimatum game. It rests on her opportunity to credibly make a *take-it-or-leave-it offer* or, equivalently, to irrevocably *commit* herself to asking for the entire surplus.[29] The situation can change dramatically if the game has two stages as in Fig. 1.8 (b). Suppose that after having rejected an offer x_0, player 2 can make a counter-offer x_1 which is also expressed in terms of player 1's surplus share. The overall surplus division will then reflect player 2's advantage of making a credible take-it-or-leave-it offer at stage 2, while the renegotiation in stage 2 renders any commitment to a particular division by player 1 incredible.

More generally, consider the *alternating offers bargaining game with n stages* illustrated in Fig. 1.8 (c). A discrete version of this game form has first been considered by Ingolf Ståhl (1972),[30] and a similar model has also been investigated by Krelle (1976, ch. 9). As in the ultimatum game, player 1 proposes $x_0 \in X = [0,1]$ corresponding to a division $(x_0, 1 - x_0)$ in the initial period $t = 0$. Player 2 can either accept or reject 1's proposal. If he

---

[29] The importance of commitment and credibility for bargaining has early been stressed by Schelling (1960, ch. 2).

[30] Ståhl also considered infinitely many stages, but assumed that players' preferences are such that there exists a last 'relevant' period, so that backward induction could be used as in the finite case. His model differs from the one below by considering fixed period surpluses which start accruing to the players in the period of agreement and accumulate until the last period $n - 1$.

rejects, the game moves on to the second stage $t = 1$ and player 2 makes a counter-proposal $x_1 \in X$ corresponding to division $(x_1, 1 - x_1)$. Player 1 can accept, or reject and make another proposal in $t = 2$. In each period $t \in T = \{0, \ldots, n - 1\}$, player $(t \bmod 2) + 1$ is to propose $x_t \in X$ corresponding to division $(x_t, 1 - x_t)$, and player $(t + 1 \bmod 2) + 1$ responds by acceptance or rejection. The game ends when either some player has accepted a division $(x_t, 1 - x_t)$ in period $t \leq n - 1$, or when proposal x_{n-1} has been rejected, i. e. the negotiations fail. The former outcome will be denoted by the pair $(x_t; t) \in X \times T$, and the latter by $D = (0; \infty)$. Consider players' preferences over these outcomes represented by von Neumann-Morgenstern utility functions $\pi_i : (X \times T) \cup \{D\} \to \mathbb{R}$ of the following particularly simple type[31]

$$\pi_1(x, t) = \delta_1^t \, x \qquad \text{and} \qquad \pi_2(x, t) = \delta_2^t \, (1 - x) \tag{1.13}$$

for discount factors $\delta_i \in (0, 1)$, $i \in I$, which reflect players' patience. This n-stage alternating offers bargaining game will be denoted by $\Gamma^{(n, \delta_1, \delta_2)}$. Since $\Gamma^{(n, \delta_1, \delta_2)}$ has finitely many stages, the method of *backward induction* yields all subgame perfect equilibria. For illustration consider again the two-stage game with $T = \{0, 1\}$ in Fig. 1.8 (b). In period $t = 1$, the game is exactly the ultimatum game though with player 2 as proposer. It has a unique SPE corresponding to stage payoffs $(0, 1)$ in $t = 1$. Player 1's proposal in period 0 has to offer player 2 at least a payoff of δ_2 in order to make player 2 weakly prefer acceptance over rejection. Player 1's optimal proposal is therefore the division $(1 - \delta_2, \delta_2)$. In the unique SPE of the two-stage game this proposal is made by player 1 and accepted by player 2, i. e. there is immediate agreement on an efficient division of the surplus.[32] This result can be extended to the general n-stage case.

Proposition 1.2. *Given the finite set of periods* $T = \{0, \ldots, n - 1\}$, $n \geq 1$, *the unique SPE outcome of the alternating offers bargaining game with* n *stages,* $\Gamma^{(n, \delta_1, \delta_2)}$, *is the efficient outcome* $(x^*(n, \delta_1, \delta_2); 0)$ *with*

$$x^*(n, \delta_1, \delta_2) = \sum_{t=0}^{\lfloor \frac{n-1}{2} \rfloor} (\delta_1 \delta_2)^t - \delta_2 \sum_{t=0}^{\lfloor \frac{n-2}{2} \rfloor} (\delta_1 \delta_2)^t, \tag{1.14}$$

where $\lfloor y \rfloor$ *denotes the biggest integer smaller than or equal to* y.

Proof. One proceeds by complete induction. Period 2 of the game $\Gamma^{(n+2, \delta_1, \delta_2)}$ with $n + 2$ periods is strategically equivalent to period 0 of the game $\Gamma^{(n, \delta_1, \delta_2)}$

---

[31] More general preferences with respect to time and elements of the underlying bargaining set will be considered in the next section.

[32] It would also be optimal for player 2 to accept only proposals strictly better than division $(1 - \delta_2, \delta_2)$ and to make the take-it-or-leave-it offer $x_1 = 0$ in stage 1. Based on this strategy, player 1 would be better off by offering $x_0 = 1 - \delta_2 - \varepsilon$ for small $\varepsilon > 0$. However, no optimal ε exists. Thus $(1 - \delta_2; 0) \in (X \times T) \cup D$ is the unique SPE outcome.

with n periods, where in equilibrium $x^*(n, \delta_1, \delta_2)$ will be offered by player 1 and accepted by player 2 by the induction hypothesis. So in period 1 of game $\Gamma^{(n+2,\delta_1,\delta_2)}$, player 2 optimally proposes division $(\delta_1\, x^*(n, \delta_1, \delta_2), 1 - \delta_1\, x^*(n, \delta_1, \delta_2))$, making it weakly preferable for player 1 to accept. In fact, $(\delta_1\, x^*(n, \delta_1, \delta_2); 1)$ is the unique SPE outcome of the subgame of $\Gamma^{(n+2,\delta_1,\delta_2)}$ starting in period 1.[33] Reasoning backwards from this, it is optimal for player 1 to propose division $(x^*, 1 - x^*)$ with

$$x^* = 1 - \delta_2\left(1 - \delta_1\, x^*(n, \delta_1, \delta_2)\right)$$

in period 0, and for player 2 to accept – corresponding to the unique SPE outcome $(x^*; 0)$. Using (1.14), one can check that

$$x^* = 1 - \delta_2\left(1 - \delta_1 \sum_{t=0}^{\lfloor\frac{n-1}{2}\rfloor}(\delta_1\delta_2)^t + \delta_1\delta_2 \sum_{t=0}^{\lfloor\frac{n-2}{2}\rfloor}(\delta_1\delta_2)^t\right)$$

$$= 1 - \delta_2 + \sum_{t=1}^{\lfloor\frac{(n+2)-1}{2}\rfloor}(\delta_1\delta_2)^t - \delta_2 \sum_{t=1}^{\lfloor\frac{(n+2)-2}{2}\rfloor}(\delta_1\delta_2)^t$$

$$= x^*(n + 2, \delta_1, \delta_2).$$

Thus if the claim holds for $\Gamma^{(n,\delta_1,\delta_2)}$, it also holds for $\Gamma^{(n+2,\delta_1,\delta_2)}$. Since the claim has already been shown to be true for $n = 1$ and $n = 2$, it must be true for all $n \in \mathbb{N}$. $\square$

Differentiating (1.14) with respect to δ_1 and δ_2 yields the following intuitive result:

Corollary 1.1. *Player 1's SPE payoff in $\Gamma^{(n,\delta_1,\delta_2)}$, $x^*(n, \delta_1, \delta_2)$, is increasing in δ_1. Similarly, player 2's SPE payoff, $1 - x^*(n, \delta_1, \delta_2)$, is increasing in δ_2. Hence it is – ceteris paribus – beneficial for each player to be a patient bargainer.*

Considering two bargainers with identical preferences, i.e. $\delta_1 = \delta_2 = \delta \in (0, 1)$, player 1's SPE share of the surplus is

$$x^*(n, \delta, \delta) = \sum_{t=0}^{n-1}(-\delta)^t$$

by (1.14). This sum of the first $n - 1$ terms of a geometric series can also be written as follows:

$$x^*(n, \delta, \delta) = \frac{1}{2} + \underbrace{\frac{1}{1+\delta} - \frac{1}{2}}_{g(\delta)=} + \underbrace{\left(-\frac{(-\delta)^n}{1+\delta}\right)}_{h(\delta,n)=}. \tag{1.15}$$

---

[33] An argument similar to that in fn. 32 establishes uniqueness.

The decomposition in (1.15) helps to illustrate two procedural aspects of *bargaining power* which typically produce an outcome different from the equal surplus division $(\frac{1}{2}, \frac{1}{2})$. First, player 1 can benefit from the fact that any proposal she makes in the initial stage 0 involves no discounting if player 2 accepts. The term $g(\delta) > 0$ can be interpreted as a measure of this *first mover advantage*. It is based on players' impatience and vanishes for $\delta \to 1$.[34] Second, as seen above, the player who is the proposer of the ultimatum game played in the final period $t = n - 1$ of $\Gamma^{(n,\delta,\delta)}$ has a *last mover advantage* due to perfect commitment when making a take-it-or-leave-it offer. This advantage is reflected by $h(\delta, n)$.[35] It is more pronounced the more patient both players are. Last mover advantage is weaker the farther away is the last period. In fact, $h(\delta, n)$ approaches zero for $n \to \infty$.

1.3.2 Rubinstein's Alternating Offers Model

The preceding section has investigated an explicit and not unrealistic bargaining process with finitely many stages. The corresponding non-cooperative game $\Gamma^{(n,\delta_1,\delta_2)}$ has a unique division $(x^*(n, \delta_1, \delta_2), 1 - x^*(n, \delta_1, \delta_2))$ which can be supported by a SPE strategy profile. The driving force behind this strong result is players' impatience to reach an agreement and one player's opportunity to make a take-it-or-leave-it offer in the final period – reflected in a superposition of first and last mover advantages. In practice, some institutionalized negotiations will have a given number n of stages that is known in advance. For most cases, however, it seems somewhat arbitrary to assume any particular such number. It is more natural to allow negotiations to – at least in principle – go on forever. This defines an *alternating offers bargaining game with infinitely many stages* which has exactly the same move structure as depicted in Fig. 1.8 (c) for the n-stage case but without a last period. The investigation of such bargaining games with infinite horizon has been pioneered by Ariel Rubinstein (1982). He has shown that the strong result of a unique SPE prediction can be generalized from the finite case to that of infinitely many stages. The following exposition will be based on later presentations of his model, mainly Osborne and Rubinstein (1990), but also Rubinstein (1987), and Binmore, Osborne, and Rubinstein (1992), all of which incorporate the elegant proof of Rubinstein's main result suggested by Shaked and Sutton (1984).

---

[34] The limit $\delta \to 1$ reflects that players become practically indifferent between outcomes $(x;t)$ and $(x;t+1)$. This can either be caused by a shrinking period length and fixed impatience per unit of real time. Or, the period length is considered fixed and players approach an ideal patience.

[35] Note that $h(\delta, n)$ is positive for odd n, i.e. increases player 1's share to the detriment of player 2, and negative for even n corresponding to the case when player 2 makes the final offer.

Crucial elements in any strategic theory of sequential bargaining are players' preferences with respect to combinations of *agreements* $x \in X$ and *agreement times* $t \in T$. This is nicely stated by Cross (1965, p. 72):

> If it did not matter when people agreed, it would not matter whether or not they agreed at all.

In above alternating offers model with a finite time horizon, a particularly simple type of preferences has been assumed (see (1.13) on p. 28). In the following, more general preferences will be allowed. The restriction to divisions $(x, 1 - x)$ of a constant surplus – corresponding to offers $x \in X = [0, 1]$ – then means little loss of generality. Namely, x and $1 - x$ can be used to parameterize all sorts of efficient forms of cooperation.[36] Thus, a rich variety of bargaining situations is covered by the following model, not only the simple division of a sum of money.

The *outcome* of an alternating offers bargaining game with infinite horizon is either a combination $(x; t) \in X \times T$ denoting division $(x, 1 - x)$ and agreement time $t \in T = \{0, 1, 2, \ldots\}$.[37] Or the outcome is alternatively an infinite sequence $(x_0, x_1, \ldots)$ of offers and counter-offers $x_t \in X$ without agreement, denoted by $D = (0; \infty)$. The assumption that the *set of outcomes* is $X \times T \cup \{D\}$ is an important restriction. In particular, players are not allowed to have preferences concerning the sequence of offers and counter-offers, $(x_0, x_1, \ldots, x_{t-1})$, before a division $(x_t, 1 - x_t)$ is eventually agreed on in period $t > 0$.[38] Moreover, only deterministic outcomes will be considered – meaning that neither player can randomize his actions.

Each player $i = 1, 2$ is assumed to have preferences over $X \times T \cup \{D\}$ that are represented by a complete, transitive, and reflexive relation $\succsim_i$.[39] The following additional properties are imposed on $\succsim_i$ for each $i = 1, 2$:

Incentive to agree (A1):
$(\forall (x; t) \in X \times T): (x; t) \succsim_i D.$

Desirable surplus (A2):
$(\forall t \in T)(\forall x, y \in X): \{x > y \Longleftrightarrow (x; t) \succ_1 (y; t) \land (x; t) \prec_2 (y; t)\}.$

Impatience (A3):
$(\forall x \in X)(\forall s, t \in T): \{t < s \Longleftrightarrow (x; t) \succ_i (x; s)\}.$

Continuity (A4):
The sets $\{(x; t) \in X \times T: (x; t) \succsim_i (y; s)\}$ *and* $\{(x; t) \in X \times T: (x; t) \precsim_i (y; s)\}$ *are closed in the product topology on* $X \times T$ *for any* $(y; s) \in X \times T$.

---

[36] Note, however, the continuity assumption (A4) which will be made.

[37] The length τ of one period of time can be chosen arbitrarily. For simplicity, the case of $\tau = 1$ is considered here.

[38] Such preferences would be incompatible with axiom (A5). Existence of a SPE would no longer be guaranteed.

[39] Cf. p. 153, n. 1, for the derived relations. Note that $\succsim_i$ will be restricted to deterministic outcomes. Hence the payoff or utility functions derived below will *not* be of the von Neumann-Morgenstern type.

Stationarity (A5):
$(\forall t \in T)\,(\forall x, y \in X): \left\{(x;t) \succ_i (y;t+1) \Longleftrightarrow (x;0) \succ_i (y;1)\right\}.$

(A1) formalizes that any agreement reached in finite time is regarded at least as good as perpetual disagreement by both players. (A2)–(A5) concern preferences on $X \times T$ and will allow particular utility representations of $\succsim_i$ on $X \times T$. (A2) and (A3) speak for themselves. (A4) is a standard technical assumption in microeconomic models; it is needed here for more than just convenience (cf. the discrete case discussed on p. 43). As investigated in detail by Fishburn and Rubinstein (1982), $\succsim_1$ satisfies (A2)–(A4) if and only if it can be represented by a continuous utility function $\pi_1 \colon X \times T \to \mathbb{R}$ where $\pi_1(x, t)$ is strictly increasing in x, and strictly decreasing in t when $x \neq 0$. An analogous statement holds for $\succsim_2$ and π_2. Finally, (A5) makes the simplification that strict preference concerning two time-dependent outcomes – and hence weak preference and indifference – depends only on the distance between the two periods and not the periods themselves. Given (A2)–(A4), the stationarity assumption (A5) allows to separate the influence of timing and actual surplus division in players' utility functions. Namely, for any fixed $\delta_1 \in (0, 1)$ there is a continuous strictly increasing function $u_1 \colon X \to \mathbb{R}$ such that $\pi_1(x, t) = \delta_1^t u_1(x)$ represents $\succsim_1$. Similarly, some u_2 yields a separable utility function π_2 representing $\succsim_2$ given an arbitrary $\delta_2 \in (0, 1)$. Note that u_i need not be concave. Moreover, given two separable utility functions π_i and π_i' of this type, $\delta_i > \delta_i'$ implies that π_i represents more patient preferences than π_i' only if the functions u_i and u_i' are identical.

(A2)–(A4) imply that the following functions $v_i \colon X \times T \to [0, 1]$ with

$$v_i(x, t) = \begin{cases} y; & (y; 0) \sim_i (x; t) \\ 0; & (\forall y \in X): (y; 0) \succ_i (x; t) \text{ for } i = 1 \\ 1; & (\forall y \in X): (y; 0) \succ_i (x; t) \text{ for } i = 2 \end{cases}$$

are well-defined for $i = 1, 2$. $v_i(x, t)$ will be referred to as the *present value* of $(x; t)$ for player i.[40] Using (A2)–(A4), $v_i(\cdot, t)$ can be shown to be continuous for any given $t \in T$. Moreover, $v_1(x, t)$ is weakly increasing in x, and strictly increasing in x for $v_1(x, t) > 0$. Also, $v_1(x, t) \leq x$ for any $(x; t) \in X \times T$, with strict inequality if x and t are both strictly positive. Analogously, v_2 is decreasing in x and satisfies opposite inequalities.

The following is a technical requirement on the present value functions v_1 and v_2 defined by $\succsim_1$ and $\succsim_2$:

Unique intersection (A6):
The equations

$$y = v_1(x, 1) \qquad and \qquad x = v_2(y, 1) \tag{1.16}$$

---

[40] This only loosely relates to the terminology in finance. The case of $v_i(x, t) = 0$ (or 1) for $x > 0$ ($x < 1$) applies e.g. when a fixed bargaining cost per round implies that even a zero share of the surplus in the initial stage – corresponding to $y = 0$ for player 1 or $y = 1$ for player 2 – is better than share x or $1 - x$, respectively, in period t.

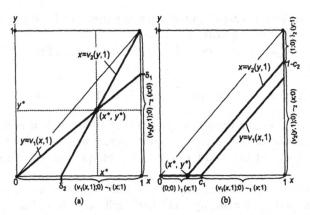

Fig. 1.9. Illustration of assumption (A6), using preferences with constant discount rates δ_i and preferences with constant costs of delay $c_1 \neq c_2$

have a unique solution $(x^*, y^*) \in X \times X$.

The case $x^* = y^* = 0$ is impossible, so that $x^* > y^*$ must hold because of $y^* = v_1(x^*, 1) < x^*$. Two examples of value functions v_1 and v_2 with a unique intersection (x^*, y^*) (necessarily below the 45° line) are given in Fig. 1.9. Equation (1.16) requires that player 1 weakly prefers to accept y^* now rather than to agree on x^* next period. Similarly, player 2 weakly prefers accepting x^* to waiting for y^* with one period of delay. Agreement pairs (x^*, y^*) with this *weak indifference property* are crucial in many variations of Rubinstein's alternating offers model,[41] and even if x^* and y^* are not uniquely defined.

(A6) does not seem a very natural requirement. However, it can be shown to hold if the following more intuitive property is satisfied in addition to (A1)–(A5) (see e.g. Osborne and Rubinstein 1990, pp. 35ff):

Strictly increasing compensation for delay (A6′):
The difference $x - v_1(x, 1)$ *and the difference* $v_2(x, 1) - x$ *are strictly increasing in* x.

(A6′), in turn, is true (as well as (A2)–(A5)) for instance if

$$\pi_1(x, t) = \delta_1^t u_1(x) \qquad \text{and} \qquad \pi_2(x, t) = \delta_2^t u_2(1 - x)$$

for weakly concave and strictly increasing functions u_i and $\delta_i \in (0, 1)$. The linear case of $u_i(x) = x$ describes *time preferences with a constant discount*

---

[41] In other settings, (x^*, y^*) can, for example, denote a pair of lotteries if the role of being the proposer is randomly assigned in each period – modelling a more symmetric bargaining process. Also, the present values in (1.16) can be considered for distinct time intervals of length τ^1 and τ^2, e.g. if players make counterproposals with different speed (see e.g. Osborne and Rubinstein 1990, pp. 53ff). Other variations of Rubinsteins's original model can be found in Binmore (1987b) and Muthoo (1999).

rate. It is shown in Fig. 1.9 (a). In this case, the discount factors δ_i unequivocally describe players' impatience.

Another very natural type of preferences is depicted in Fig. 1.9 (b) and is represented by

$$\pi_1(x,t) = x - c_1 t \quad \text{and} \quad \pi_2(x,t) = (1-x) - c_2 t \tag{1.17}$$

for $c_i \in [0,1)$. (1.17) describes *time preferences with constant cost of delay* $c_i \geq 0$ for each player. For $c_1, c_2 > 0$, time preferences with constant cost of delay satisfy (A1)–(A5), and for $c_1 \neq c_2$ the unique intersection property (A6) holds.

All *alternating offers bargaining games with infinitely many stages* for which players' preferences satisfy (A1)–(A6) will be collected in the set $\mathcal{R}$. An element $\Gamma^{(\infty, \succsim_1, \succsim_2)} \in \mathcal{R}$ will be referred to as a *Rubinstein bargaining game.*

Let X^t denote the set of all sequences $(x_0, x_1, \ldots, x_{t-1})$ of t offers and counter-offers in $X = [0,1]$. Elements of X^t describe *bargaining histories* of length $t \in T = \{0, 1, 2, \ldots\}$. A strategy for player 1 in a game $\Gamma^{(\infty, \succsim_1, \succsim_2)} \in \mathcal{R}$ is a collection $s_1 = \{s_1^t\}_{t \in T}$ of functions that map a given bargaining history to an action.[42] If t is even, player 1 makes a proposal $x_t = s_1^t(x_0, \ldots, x_{t-1})$ based on history $(x_0, \ldots, x_{t-1})$; hence $s_1^t : X^t \to X$. Similarly, if t is odd, player 1 responds to the proposal $x_t \in X$ of player 2; hence $s_1^t : X^{t+1} \to \{0,1\}$. Player 2's strategies are analogously collections $s_2 = \{s_2^t\}_{t \in T}$ with $s_2^t : X^t \to X$ if t is odd and $s_2^t : X^{t+1} \to \{0,1\}$ if t is even. Note that this formulation allows players to base their action in period t on the entire bargaining history before t. Thus strategies are not restricted to be *stationary* in the sense of always specifying the same response to any particular offer $x \in X$.

As already observed for the ultimatum game, the Nash equilibrium concept is extremely weak in the context of sequential bargaining. Any division $(\bar{x}, 1 - \bar{x})$ with $\bar{x} \in X$ can be agreed upon in the initial stage $t = 0$ if agents play the Nash equilibrium (s_1, s_2) with

$$s_1^t(x_0, \ldots, x_{t-1}) = \bar{x} \text{ for all } (x_0, \ldots, x_{t-1}) \in X^t \text{ if } t \in T \text{ is even, and}$$

$$s_1^t(x_0, \ldots, x_t) = \begin{cases} 0; & x_t < \bar{x} \\ 1; & x_t \geq \bar{x} \end{cases} \quad \text{if } t \in T \text{ is odd,}$$

and an analogous strategy s_2 such that player 2 always proposes $\bar{x}$, rejects offers $x > \bar{x}$, and accepts $x \leq \bar{x}$. If players are patient enough, many divisions $(\bar{x}, 1 - \bar{x})$ can be induced by NE strategies even in later periods $t > 0$ – implying an inefficient waste of surplus. However, given a NE that induces e. g. the division $(1,0)$ in $t = 0$, the corresponding strategy s_1 is based on an *incredible threat*: Should player 2 reject $x_0 = 1$ and propose $x_1 = v_1(1,1) + \varepsilon < 1$ for $\varepsilon > 0$ in $t = 1$, player 1 strictly prefers to accept rather than to carry

---

[42] X^0 is the unspecified *initial history*, and s_1^0 is simply an element of X.

out her threat of rejecting x_1 and proposing $x_2 = 1$ again in the next period. Therefore, (s_1, s_2) induces no NE in the subgame which starts with player 1 responding to x_1. Strategy profiles which involve such sequentially irrational behaviour are ruled out by Selten's (1965) subgame perfectness refinement. When the SPE concept is applied, a very strong result holds for all Rubinstein bargaining games.

In particular, consider the unique pair $(x^*, y^*) \in X \times X$ introduced in (A6) and the strategy profile (s_1^*, s_2^*) with, first,

$$s_1^{*t}(x_0, \ldots, x_{t-1}) = x^* \text{ for all } (x_0, \ldots, x_{t-1}) \in X^t \text{ if } t \in T \text{ is even, and}$$

$$s_1^{*t}(x_0, \ldots, x_t) = \begin{cases} 0; \ x_t < y^* \\ 1; \ x_t \geq y^* \end{cases} \qquad \text{if } t \in T \text{ is odd,}$$

meaning that player 1 always proposes x^* when it is her turn, and accepts an offer y of player 2 if and only if $y \geq y^*$. Second, s_2^* is the analogous strategy for player 2, specifying that he always proposes y^*, rejects offers x from player 1 if $x > x^*$, and accepts $x \leq x^*$. Based on this definition, the following holds:

Theorem 1.3. (Rubinstein) *For any Rubinstein bargaining game* $\Gamma^{(\infty, \succsim_1, \succsim_2)} \in \mathcal{R}$

i) strategy profile (s_1^*, s_2^*) *is a subgame perfect equilibrium of* $\Gamma^{(\infty, \succsim_1, \succsim_2)}$,
ii) moreover, (s_1^*, s_2^*) *is the unique SPE of* $\Gamma^{(\infty, \succsim_1, \succsim_2)}$.

Thus in the unique SPE outcome of a Rubinstein bargaining game player 1 proposes x^* *as defined by (A6) in period* $t = 0$, *and player 2 immediately accepts.*

Proof. Let the sets $\mathcal{R}_1$ and $\mathcal{R}_2$ denote the collections of the subgames of $\Gamma^{(\infty, \succsim_1, \succsim_2)}$ starting with a proposal of player 1 and 2, respectively. By the stationarity assumption (A5), all subgames $\Gamma_i \in \mathcal{R}_i$ are strategically equivalent.

i) Consider a subgame $\Gamma_1 \in \mathcal{R}_1$ that starts in an arbitrary even period t_1 of $\Gamma^{(\infty, \succsim_1, \succsim_2)}$. In Γ_1, t_1 is the initial period and can be renamed $t = 0$. If player 1 proposes x^* as specified by s_1^*, player 2 will – using strategy s_2^* – accept, yielding the outcome $(x^*; 0)$. Suppose that player 1 deviates and proposes $x' \neq x^*$. First, assume $x' < x^*$. This proposal will be accepted by 2, yielding the outcome $(x'; 0)$ which by (A2) is, however, worse for player 1 than $(x^*; 0)$. Second, assume $x' > x^*$. Then x' will be rejected by s_2^*, and the possible outcomes of the game are either

    a) $(y^*; t)$ for $t \geq 1$ if player 1 directly afterwards or later on accepts a counter-offer specified by s_2^*, or

    b) $(x''; t)$ for $x'' \leq x^*$ and $t \geq 2$ if player 1 rejects y^*, and directly afterwards or later makes a new offer x'' which is accepted by s_2^*, or

    c) perpetual disagreement D.

By $x^* > y^*$, (A2), and (A3), option (a) is worse than $(x^*; 0)$. By (A3), option (b) is worse than $(x^*; 0)$. Finally, by (A1), option (c) is not preferred to $(x^*; 0)$ either. So given that player 2 uses strategy s_2^*, it is optimal for player 1 to propose x^* in the first period of Γ_1, and hence in any even period t_1 of $\Gamma^{(\infty, \succsim_1, \succsim_2)}$.

Similarly, it is optimal for player 2 to accept any proposal $x \leq x^*$ in Γ_1 given that player 1 uses s_1^*: Acceptance yields outcome $(x^*; 0)$. Rejection, however, produces either $(y; t)$ for $y \geq y^*$ and $t \geq 1$, which by $x^* > y^*$ and (A3) is not preferred to $(x^*; 0)$, or, the disagreement outcome D, which by (A1) is never preferred. Analogous arguments show that it is optimal for player 2 to make the proposal y^* in the initial period of any subgame $\Gamma_2 \in \mathcal{R}_2$ that starts in an arbitrary odd period t_2, and that it is optimal for player 1 to immediately accept any offer y with $y \geq y^*$ in such a period t_2. It can also be checked that each s_i^* is a best reply to s_{-i}^* in each subgame of $\Gamma^{(\infty, \succsim_1, \succsim_2)}$ that starts with the acceptance or rejection decision of player i.

ii) Make the following definitions for subgames $\Gamma_i \in \mathcal{R}_i$ and $i = 1, 2$:

$$M_i = \sup \{v_i(x, t) : (x; t) \text{ is a SPE outcome of } \Gamma_i\}$$

and

$$m_i = \inf \{v_i(x, t) : (x; t) \text{ is a SPE outcome of } \Gamma_i\}.$$

By (A5), it does not matter, which subgames Γ_1 and Γ_2 are chosen in these definitions. Also, it is possible to use two arbitrary subsequent periods in the following argument – with periods 0 and 1 as the most convenient choice.

Loosely speaking, M_1 and m_2 are the present value of the best outcome that players 1 and 2, respectively, can expect in any SPE of any subgame of $\Gamma^{(\infty, \succsim_1, \succsim_2)}$ that starts with their proposal. Similarly, m_1 and M_2 is the present value of the worst such outcome. It will be shown, first, that $M_1 = x^*$ and $M_2 = y^*$.

Step 1: Suppose that in the initial period $t = 0$ of subgame $\Gamma_2 \in \mathcal{R}_2$ – corresponding to some odd period t_2 of $\Gamma^{(\infty, \succsim_1, \succsim_2)}$ – player 2 proposes x with $x > v_1(M_1, 1)$. By rejecting x, player 1 can obtain a SPE outcome with at most the present value $v_1(M_1, 1)$; hence in any SPE player 1 must accept such an offer x. This means that player 2 must even in his least desired SPE outcome get the share $1 - v_1(M_1, 1)$, implying[43]

$$M_2 \leq v_1(M_1, 1). \tag{1.18}$$

Step 2: x^* is the present value of a SPE outcome in a game $\Gamma_1 \in \mathcal{R}_1$ by part i). The best SPE outcome from player 1's point of view cannot be worse, i. e.

---

[43] Recall that proposals $x \in X$, present values of outcomes, and the respective infima and suprema, m_i and M_i, are all defined in terms of player 1's share of the surplus.

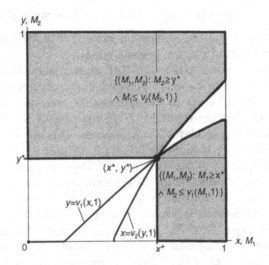

Fig. 1.10. Illustration of the proof of Theorem 1.3 ii)

$$M_1 \geq x^*. \tag{1.19}$$

All $(M_1, M_2) \in X \times X$ that satisfy (1.18) and (1.19) are shown in the lower grey area in Fig. 1.10.

Step 3: If in the initial period $t = 0$ of a subgame $\Gamma_1 \in \mathcal{R}_1$ player 2 rejects an offer x, then he is able to obtain at least the outcome $(M_2; 1)$ in the next period. This implies that player 1 can ask at most for $v_2(M_2, 1)$ in a SPE in $t = 0$ if her proposal is to be accepted by player 2. Since any outcome which does not involve immediate acceptance by player 2 has at most the present value $v_1(M_2, 1) \leq M_2 \leq v_2(M_2, 1)$, any SPE of Γ_1 has a present value for player 1 no greater than $v_2(M_2, 1)$, implying

$$M_1 \leq v_2(M_2, 1). \tag{1.20}$$

Step 4: y^* is the present value of a SPE outcome in a game $\Gamma_2 \in \mathcal{R}_2$ by part i). The worst SPE outcome from player 2's point of view cannot be better, i.e.

$$M_2 \geq y^*. \tag{1.21}$$

All $(M_1, M_2) \in X \times X$ that satisfy (1.20) and (1.21) are shown in the upper grey area in Fig. 1.10. Hence, combining (1.18)–(1.21) yields $(x^*, y^*) = (M_1, M_2)$. Carrying out steps 1–4 once again but assuming that player 1 proposes x with $x < v_2(m_2, 1)$ in step 1, considering what player 1 can obtain by rejecting an offer x in step 3, and stating the reverse inequalities for m_i in steps 2 and 4, one similarly obtains that $(x^*, y^*) = (m_1, m_2)$. Thus

$$M_1 = m_1 = x^* \qquad \text{and} \qquad M_2 = m_2 = y^*. \tag{1.22}$$

This establishes that the Rubinstein game $\Gamma^{(\infty, \succsim_1, \succsim_2)}$ has a unique SPE outcome in terms of players' present values.

The result carries over to the set of SPE itself. Assume that a SPE (s_1', s_2') of $\Gamma^{(\infty, \succsim_1, \succsim_2)}$ exists in which player 2 does not immediately agree to player 1's offer $s_1'^0 \in X$. In the subgame Γ_2 that starts with player 2's counter-offer after the rejection of $s_1'^0$, the present value of the SPE of Γ_2 is y^* according to player 2's preferences. This implies that player 1's present value in Γ_2 induced by (s_1', s_2') is at most y^*. Player 1's present value of getting y^* in period 1 of $\Gamma^{(\infty, \succsim_1, \succsim_2)}$ is $v_1(y^*, 1) \leq y^* < x^*$. That the present value of the SPE outcome induced by (s_1', s_2') should be smaller than x^* contradicts $M_1 = m_1 = x^*$. Therefore there must be immediate agreement on player 1's offer x^* in any SPE of $\Gamma^{(\infty, \succsim_1, \succsim_2)}$, and in fact every $\Gamma_1 \in \mathcal{R}_1$. A similar argument establishes that there must be immediate agreement on player 2's offer y^* in any SPE of every $\Gamma_2 \in \mathcal{R}_2$. This means that player 1 always proposes x^* and player 2 always proposes y^* when it is 1's and 2's turn, respectively.

It remains to check that player 1 rejects an offer x if and only if $x < y^*$, and that player 2 rejects an offer x if and only if $x > x^*$ in any SPE. The if-part must be true to make x^* and y^*, respectively, an optimal proposal. Concerning the only- if-part, note that a rejection of proposals strictly greater than y^* or smaller than x^*, respectively, would be suboptimal given (1.22) and (A3).

$\square$

Note that for part i), only the existence requirement of (A6) is used. If (1.16) has multiple solutions (x^*, y^*), each such pair gives rise to a SPE. Then, however, subgame perfectness no longer implies immediate agreement, i.e. surplus may be lost due to inefficient delay. An example of this is given for preferences with constant costs of delay $c_1 = c_2$ in Rubinstein (1982, pp. 107ff). In case of $c_1 > c_2$, it can directly by inferred from Fig. 1.9 (b) that $x^* = c_2$, i.e. player 1's share is effectively as in a two-stage alternating offers game where player 2 is the last mover. Similarly, $x^* = 1$ for $c_1 < c_2$ – with the intuition that player 2 would be the definite loser of a 'war of attrition' and cannot do better than to accept a zero share right away.

In case of preferences with constant rate of discount, (1.16) translates into

$$y = \delta_1 x \qquad \text{and} \qquad x = 1 - \delta_2(1 - y)$$

with the unique solution

$$x^*(\delta_1, \delta_2) = \frac{1 - \delta_2}{1 - \delta_1 \delta_2} \qquad \text{and} \qquad y^*(\delta_1, \delta_2) = \frac{\delta_1(1 - \delta_2)}{1 - \delta_1 \delta_2}. \tag{1.23}$$

This corresponds to agreement on the surplus division $\left(\frac{1 - \delta_2}{1 - \delta_1 \delta_2}, \frac{\delta_2(1 - \delta_1)}{1 - \delta_1 \delta_2} \right)$ in $t = 0$ as the unique SPE outcome. It is noteworthy that this is – non-trivially –

also the limit of the SPE division derived for the alternating offers game with finite horizon, $\Gamma^{(n,\delta_1,\delta_2)}$, in Proposition 1.2 (p. 28), i. e.

$$\lim_{n \to \infty} x^*(n, \delta_1, \delta_2) = x^*(\delta_1, \delta_2).$$

Assuming players with identical patience $\delta_1 = \delta_2 = \delta \in (0,1)$ in (1.23), player 1 gets a share of $\frac{1}{1+\delta} > \frac{1}{2}$. Thus *first mover advantage* exists also in bargaining with infinite horizon. This holds for identical players with more general preferences, too: If $(x;t) \succsim_1 (y;s) \iff (1-x;t) \succsim_2 (1-y;s)$ holds for all $x, y \in X$ and $s, t \in T$, then $v_2(1-x, t) = 1 - v_1(x, t)$. Hence, (1.16) is solved for $x^* = 1 - y^*$; with $x^* > y^*$ this means $x^* > \frac{1}{2}$.

A noticeable advantage for the first mover does not seem an obvious feature of bargaining in practice. Note, however, that player 1's advantage diminishes as the discount factor for one period length of bargaining increases. For $\delta \to 1$ in above symmetric case, the SPE division approaches an equal share of surplus.[44]

Player's relative patience, in contrast, seems to matter in practice and certainly does in theory. In (1.23), player 1's share increases in δ_1 and decreases in δ_2. More generally, preferences $\succsim_1'$ are *more patient* than preferences $\succsim_1$ if the inequality $v_1'(x, 1) \geq v_1(x, 1)$ holds, and is strict for some $x \in X$. This implies that the graph of v_1' lies (weakly) above that of v_1 in Fig. 1.9 so that $x^{*'} \geq x^*$, i. e. it is ceteris paribus beneficial to be more patient.

Now consider players with preferences that are represented by

$$\pi_1(x,t) = \delta_1^t u_1(x) \qquad \text{and} \qquad \pi_2(x,t) = \delta_2^t u_2(1-x) \qquad (1.24)$$

for $t \in \mathbb{R}_+$, $\delta_i \in (0,1)$, and weakly concave, strictly increasing functions $u_i : X \cup \{D\} \to \mathbb{R}$ with $u_i(D) = 0$ for $i = 1, 2$. For these preferences, present values v_i are defined by indifference and not only weak preference, i. e. $y = v_i(x, t) \iff (y; 0) \sim_i (x; t)$. Generally, let each period of $\Gamma^{(\infty, \succsim_1, \succsim_2)}$ have the length $\tau > 0$ and denote solutions of (1.16) obtained for given τ by $(x^{*\tau}, y^{*\tau})$. There is a useful technical relationship between the unique SPE outcome of a Rubinstein bargaining game $\Gamma^{(\infty, \succsim_1, \succsim_2)}$ and the cooperative Nash solution of a corresponding bargaining problem. This has first been pointed out in the literature by Binmore (1987a).

Proposition 1.3. (Binmore) *Let players $i = 1, 2$ have preferences represented by π_i in (1.24). As the length $\tau > 0$ of a bargaining period in a Rubinstein game $\Gamma^{(\infty, \succsim_1, \succsim_2)} \in \mathcal{R}$ approaches 0, the SPE payoffs of the players, $\pi_1(x^{*\tau}, 0)$ and $\pi_2(x^{*\tau}, 0)$, approach the utility levels of the asymmetric Nash bargaining solution $F^{N(\alpha, \beta)}$ with bargaining powers*

---

[44] Assume a surplus of €100, a yearly interest rate of 5%, and ten seconds of delay before a counter-offer is made, and either accepted or rejected. Then the approximate discount rate is $\delta = 0.999999985$ under continuous compounding, and the resulting share of player 1 is €50.00000039.

$$\alpha = \frac{\ln \delta_2}{\ln \delta_1 + \ln \delta_2} \quad \text{and} \quad \beta = 1 - \alpha$$

of the bargaining problem $\langle U(u_1(\cdot), u_2(\cdot)), 0 \rangle$ defined by

$$U(u_1(\cdot), u_2(\cdot)) = \{(u_1, u_2) \in \mathbb{R}^2 : u_1 \le u_1(x) \wedge u_2 \le u_2(1-x) \wedge x \in X\}.$$

Proof. First, note that $\pi_1(x^{*\tau}, 0) = u_1(x^{*\tau})$ and $\pi_2(x^{*\tau}, 0) = u_2(1 - x^{*\tau})$. By (1.16),

$$(y^{*\tau}; 0) \sim_1 (x^{*\tau}; \tau) \iff u_1(y^{*\tau}) = \delta_1^\tau u_1(x^{*\tau})$$
$$(x^{*\tau}; 0) \sim_2 (y^{*\tau}; \tau) \iff u_2(1 - x^{*\tau}) = \delta_2^\tau u_2(1 - y^{*\tau}).$$

The two equations can be combined to yield

$$\left(\frac{u_1(y^{*\tau})}{u_1(x^{*\tau})}\right)^\alpha = \left(\frac{u_2(1 - x^{*\tau})}{u_2(1 - y^{*\tau})}\right)^\beta = e^{\tau / [1/\ln \delta_1 + 1/\ln \delta_2]} \qquad (1.25)$$

for any $\tau > 0$. By cross-multiplication, one obtains

$$u_1(x^{*\tau})^\alpha u_2(1 - x^{*\tau})^\beta = u_1(y^{*\tau})^\alpha u_2(1 - y^{*\tau})^\beta.$$

This means that the utility combinations $(u_1(x^{*\tau}), u_2(1 - x^{*\tau}))$ and $(u_1(y^{*\tau}), u_2(1 - y^{*\tau}))$ – corresponding to the SPE payoffs in a Rubinstein bargaining game that starts with player 1's and player 2's proposal, respectively – always have the same generalized Nash product with bargaining powers α and β. Since

$$\lim_{\tau \to 0} e^{\tau / [1/\ln \delta_1 + 1/\ln \delta_2]} = 1,$$

one obtains from (1.25) that

$$\lim_{\tau \to 0} u_1(y^{*\tau}) = \lim_{\tau \to 0} u_1(x^{*\tau}) = u_1^* \qquad (1.26)$$

and

$$\lim_{\tau \to 0} u_2(1 - y^{*\tau}) = \lim_{\tau \to 0} u_2(1 - x^{*\tau}) = u_2^*. \qquad (1.27)$$

The points $(u_1(x^{*\tau}), u_2(1 - x^{*\tau}))$ and $(u_1(y^{*\tau}), u_2(1 - y^{*\tau}))$ always lie on both the hyperbola defined by $u_1^\alpha u_2^\beta = c$, where c is their common generalized Nash product, and also the weakly concave Pareto frontier of $U(u_1(\cdot), u_2(\cdot))$, $P(U(u_1(\cdot), u_2(\cdot)))$. This is illustrated in Fig. 1.11. It can be seen that as both $(u_1(x^{*\tau}), u_2(1-x^{*\tau}))$ and $(u_1(y^{*\tau}), u_2(1-y^{*\tau}))$ approach (u_1^*, u_2^*), their common generalized Nash product increases, and in fact,

$$(u_1^*, u_2^*) = \underset{(u_1, u_2) \in P(u_1(\cdot), u_2(\cdot))}{\arg\max} u_1^\alpha u_2^\beta.$$

By Theorem 1.2, (u_1^*, u_2^*) then coincides with the payoff vector determined by the asymmetric Nash bargaining solution of $\langle U(u_1(\cdot), u_2(\cdot)), 0 \rangle$. $\qquad \square$

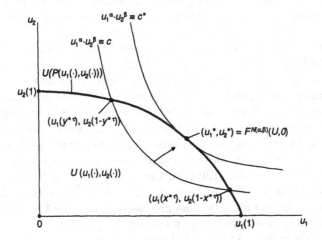

Fig. 1.11. Convergence of SPE payoffs in a Rubinstein bargaining game to the asymmetric Nash bargaining solution

Note that (1.26)–(1.27) and u_i's strict monotonicity imply that

$$\lim_{\tau \to 0} x^{*\tau} = \lim_{\tau \to 0} y^{*\tau}.$$

Hence, the first mover advantage vanishes as both players' disutility of one period of delay becomes negligible[45] – a fact that has already been observed above in a more special case (p. 30).

It is incorrect to translate Proposition 1.3 as saying that the unique SPE of a Rubinstein bargaining game converges to the Nash bargaining solution. First, the result refers only to the respective associated utility levels, and is only true when players preferences can be represented as in (1.24) with the stated restrictions. Second, utilities in both models have a different interpretation, and different properties. For the Nash solution, utility functions represent preferences over lotteries of outcomes, and are of the von Neumann-Morgenstern type. In the alternating offers model, they represent time-dependent preferences over deterministic outcomes – bearing no relation to von Neumann-Morgenstern utility.

Still, Proposition 1.3 loosely provides a non-cooperative foundation for the Nash solution and its maximization of a weighted geometric mean of players' payoff gains. It is considered an important contribution to the *Nash programme* of providing links between cooperative and non-cooperative game theory.[46] If one is not too much concerned with the subtleties of interpreting

---

[45] Again, this is equivalently reflects either more patient preferences given a fixed period length, or a shorter period length given fixed preferences.

utility functions, one can, in fact, use Proposition 1.3 to justify particular bargaining powers α and β – which have been considered above as given rather than derived from any first principles such as preferences.

Rubinstein's model of alternating offers lends itself to multiple variations, many of which are collected by Muthoo (1999). One prominent modification is the following: Consider players which are perfectly patient, i. e. value an agreed division independently of the agreement time. Assume that they face the risk that with an exogenous probability p the negotiations break down, corresponding to an 'agreement' B that is not preferred to even a zero share of the surplus by both players. This setting requires players' preference to be defined over lotteries. It is convenient to assume that they are represented by von Neumann-Morgenstern utility functions π_1 and π_2 based on weakly concave and strictly increasing functions $u_i : X \to \mathbb{R}$ which measure the utility of deterministic outcomes. The reasoning used in the proof of Theorem 1.3 can be adapted to this setting. Again the pair of agreements (x^*, y^*) in (A6) defines the unique SPE. The adapted equations are

$$u_1(y^*) = (1-p)u_1(x^*) \quad \text{and} \quad u_2(1-x^*) = (1-p)u_2(1-y^*),$$

corresponding to those for (1.24) when $\delta_1 = \delta_2 = 1 - p$. Choosing $\pi_i(B) = u_i^B = 0$, the convergence result of Proposition 1.3 holds – with the same von Neumann-Morgenstern utilities as in the original Nash solution. Thus a non-cooperative foundation of the latter is provided without above interpretational caveats.[46]

Another important set of variations of the original alternating offers model allows one or both players to choose to leave the game after they have or their opponent has rejected a proposal (cf. e. g. Sutton 1986 or Muthoo 1999, ch. 5). In case that player i has *opted out*, the outcome O_i results, corresponding to (discounted) utility levels $u^{O_i} = (u_1^{O_i}, u_2^{O_i})$. The intuition that a better *outside option* $u_i^{O_i}$ will ceteris paribus always result in an improved SPE payoff turns out to be wrong. Loosely speaking, player i's outside option does not have any effect on i's equilibrium payoff at all unless it exceeds i's SPE payoff in the Rubinstein bargaining game without outside option. Only then is it a credible threat for player i to opt out. In this case, player $-i$ must make a proposal that leaves i weakly better off than waiting for his outside option if $-i$ wants to avoid $u_{-i}^{O_i}$. This finding of an outside option having either no or a very drastic effect on the SPE division is also referred to as the *outside option principle*. The result is, however, not very robust to the exact

---

[46] Nash (1953) himself suggested a particular non-cooperative foundation of his solution based on a single-stage game in which players simultaneously make payoff 'demands' (see Sect. 1.4, p. 54). This model is discussed in detail by Binmore (1987a).

[47] Compare Binmore, Rubinstein, and Wolinsky (1986) for a more comprehensive treatment of this and the previous strategic foundation of the Nash solution. They derive useful guidelines for the application of the cooperative Nash bargaining solution in economic models based on the non-cooperative approach.

timing of players' opportunities to opt out (see, for example, Osborne and Rubinstein 1990, pp. 54–63). In particular, a great multiplicity of subgame perfect equilibria can arise from the introduction of outside options (nicely shown by Ponsati and Sakovics 1998).

The distinct possibilities of perpetual bargaining without agreement, D, an exogenous breakdown of negotiations B, or players' having outside options O_i suggest different possible choices for the disagreement point u^D entering the (asymmetric) Nash bargaining solution. To equate u^D with players' respective outside option payoffs is typically wrong by the outside option principle. Either the point $u^O = (u_1^{O_1}, u_2^{O_2})$ is irrelevant, or it is a binding restriction on the set of individually rational outcomes even as the period length τ becomes negligible. Thus, one correctly imposes $u \geq u^O$ as an additional constraint for the maximization of the Nash product. If players' primary incentive to reach an agreement rather sooner than later is due to their impatience, then the utility levels associated with outcome D are the correct choice for u^D (Proposition 1.3). When the risk of an exogenously caused breakdown of negotiations is the driving force towards agreement, then u^D should be chosen to correspond to B. A model that explicitly combines impatience, outside options, exogenous breakdown risk, and even *inside options*, which reflect players' payoff opportunities whilst they temporarily disagree, is considered by Muthoo (1999, ch. 6).

Rubinstein's (1982) bargaining model is widely appraised because it combines a somewhat realistic negotiation process – allowing, at least, for the interchange of offers and counter-offers – with a clear-cut prediction. Unfortunately, the clear-cut prediction based on a unique SPE is not too robust. The case of multiple SPE when players have outside options has already been mentioned. Van Damme, Selten, and Winter (1990) show that the uniqueness result of Theorem 1.3 does not hold if players cannot choose agreements from a continuous interval $X = [0, 1]$ (or their preferences violate (A4)), either. Suppose that players have preferences with constant discount rate as in (1.24), but can make proposals only in multiples of a smallest – e. g. monetary – unit ξ. Any outcome $(\bar{x}, 0)$ with $\bar{x} > x^*$, for example, is not supported by a SPE in the original model. This is so, because player 2 could gain from rejecting 1's offer $x_0 = \bar{x}$ and from proposing $x_1 = \bar{x} - \varepsilon$ in $t = 1$, on which player 1 could not improve by proposing $x_2 = \bar{x}$ if $\varepsilon > 0$ is sufficiently small. If, however, a smallest unit ξ prevents player 2 from making a counter-offer x_1 such that player 1 would be irrational to reject it, then $(x_0; 0)$ can be supported. This is accomplished by a SPE in which player 1 always proposes $\bar{x}$ and accepts a proposal x if and only if $x \geq \bar{x}$, and player 2 behaves analogously. Van Damme et al. (1990) show that for any $\xi > 0$, any efficient agreement can be supported by a SPE as long as players' discount factors $\delta_i \in (0, 1)$ are sufficiently large or, equivalently, the period length τ is sufficiently small. Based on these SPE-inducing efficient outcomes, equilibria inducing inefficient delay can be defined; even perpetual disagreement $(0; \infty)$ can be supported as a SPE.

An alternative modification, with multiple and also inefficient SPE, allows for additional moves between the rejection decision and a counter-proposal. As a realistic example, Haller and Holden (1990) and, in more detail, Fernandez and Glazer (1991) consider a trade union which has to decide whether to strike or not after each rejection.

A generalization of Rubinstein bargaining games to $n \geq 3$ players is possible in many different ways. Unfortunately, a unique SPE is a feature of few of them. A quite natural generalization suggested by Shaked is a negotiation process which starts by player 1 proposing $x^0 = (x_1^0, x_2^0, x_3^0)$ with $\sum x_i = 1$ in period $t = 0$. Players 2 and 3 sequentially (or, with little difference, simultaneously) accept or reject. If one of them has rejected, player 2 proposes x^1 in period $t = 1$. Players 1 and 3 accept or reject, and if one of them has rejected, player 3 proposes x^3 in $t = 3$, etc. If the players are reasonably patient, agreement on any efficient division (x_1, x_2, x_3) can be supported as a SPE (cf. e. g. Osborne and Rubinstein 1990, pp. 63–65). Typically, unique predictions for n-person alternating offers bargaining games are possible only by restricting attention to SPE in which players use *stationary strategies*, i. e. do not condition their actions on the entire bargaining history. Alternatively, the requirement that all players must agree on a proposed division can be given up to yield unique SPE predictions (see Kultti 1994, for example). But when unanimity is not required, different sets of accepting players are realistically associated with a different surplus. This case is treated by non-cooperative models of *coalition formation*, e. g. Krishna and Serrano (1995), Hart and Mas-Colell (1996), and Okada (1996).

1.3.3 Strategic Bargaining with Incomplete Information

In Rubinstein's (1982) alternating offers bargaining model, players have the opportunity to engage in a sequence of offers and counter-offers. However, in the unique SPE they agree efficiently on division $(x^*, 1 - x^*)$ right in the initial period. Particular variations of the model have multiple SPE, in some of which rejections and counter-offers can be observed. But an actual process of haggling and bidding until eventually hands are shook remains the exception – though this is commonly associated with the term 'bargaining.' In this context, Nash (1953, p. 138) observes:

> With people who are sufficiently intelligent and rational there should not be any question of 'bargaining ability,' a term which suggests something like skill in duping the other fellow. The usual haggling process is based on imperfect information, the hagglers trying to propagandize each other into misconceptions of the utilities involved.

In fact, sequences of offers and rejections are a common feature of strategic bargaining models with incomplete information about one or both bargainers' preferences. Unfortunately, the judgement of Fudenberg and Tirole (1991, p. 399) is still rather accurate:

> The theory of bargaining under incomplete information is currently more a series of examples than a coherent set of results. This is unfortunate because bargaining derives much of its interest from incomplete information.

But recall that results for modifications of the Rubinstein bargaining game with complete information have not been perfectly coherent either. Rather, different models of bargaining with complete or incomplete information highlight complementing and opposing factors actuating bargaining results with a different focus.

Models of bargaining with incomplete information are technically a lot more involved than those assuming complete information. However, a rather simple bilateral monopoly model of 2-stage bargaining with offers only from player 1 suffices to point out some typical features. Quite detailed introductions to bargaining models with incomplete information are given by Fudenberg and Tirole (1991, ch. 10) and Kennan and Wilson (1993).

Suppose that players 1 and 2 have preferences with constant discount rate, i. e.

$$\pi_1(x,t) = \delta_1^t x \qquad \text{and} \qquad \pi_2(\theta,x,t) = \delta_2^t(\theta - x),$$

for $\delta_i \in (0,1)$. θ represents player 2's valuation of a good that can be sold to him by player 1 at price $x \in X = [0,1]$. θ – also referred to as player 2's *type* – is private information, i. e. only player 2 observes the realization of a random variable $\tilde{\theta}$. It is common knowledge that player 1's reservation value for the good is zero and that $\tilde{\theta}$ is uniformly distributed on $\Theta = [0,1]$.

The game proceeds as follows (see Fig. 1.12): In period $t = 0$, Nature draws θ from Θ. Player 1 then offers a price $x_0 \in X$. Based on his observation of θ, player 2 either accepts – an outcome denoted by $(\theta, x_0; 0)$ – or rejects. In the latter case, the game moves on to period $t = 1$, and player 1 offers a second price $x_1 \in X$. If this offer is accepted by 2, payoffs are $(\delta_1 x_1, \delta_2(\theta - x_1))$, and otherwise the utility vector $(0,0)$ results.

A strategy for player 1 is a collection $s_1 = \{s_1^0, s_1^1\}$ of functions, where s_1^0 is degenerated and simply an element of X. s_1^1 maps 1's rejected initial offer x_0 to a new offer $s_1^1(x_0) \in X$. In addition, player 1 has adapting *beliefs* about player 2's valuation. The initial beliefs μ_1^{*0} are correct in the sense of corresponding exactly to the true uniform distribution of $\tilde{\theta}$. Beliefs $\mu_1^1(x_0)$ held in period $t = 1$ account for the fact that player 2 has rejected offer $x_0 \in X$ and are deduced by Bayesian updating. Let $F_\theta^0(y)$ and $F_\theta^1(y|x_0)$ denote the cumulative distribution functions of θ corresponding to μ_1^{*0} and $\mu_1^1(x_0)$, respectively.

Player 2's strategy is a collection $s_2 = \{s_2^0, s_2^1\}$, where $s_2^0: \Theta \times X \to \{0,1\}$ defines whether offer x_0 is rejected or accepted by player 2 based on his type $\theta \in \Theta$. Similarly, $s_2^1: \Theta \times X \times X \to \{0,1\}$ is 2's decision in $t = 1$. Player 2 has only trivial beliefs μ_2^*.

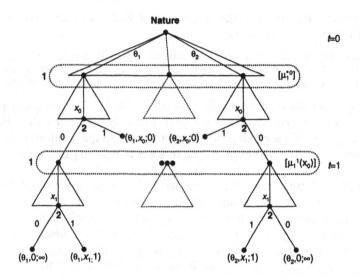

Fig. 1.12. 2-stage bargaining game with incomplete information

The game can be solved for perfect Bayesian equilibria (PBE) by back-ward induction: Player 2 will optimally accept an offer $x_1 \in X$ if and only if $x_1 \leq \theta$,[48] i.e.

$$s_2^{*1}(\theta, x_0, x_1) = \begin{cases} 1; x_1 \leq \theta \\ 0; x_1 > \theta. \end{cases}$$

Anticipating some offer $x_1 \in X$ for $t = 1$, player 2's choice in $t = 0$ is between the payoffs $\theta - x_0$ and $\max\{\delta_2(\theta - x_1), 0\}$. In $t = 0$, player 2's optimal decision is to accept x_0 if and only if the former payoff is no smaller than the latter, which is equivalent to

$$\theta \geq \theta^* = \max\left\{\frac{x_0 - \delta_2 x_1}{1 - \delta_2}, x_0\right\}. \tag{1.28}$$

So given arbitrary $x_0, x_1 \in X$, there exists a threshold valuation θ^* such that all types above θ^* accept x_0 and all types below θ^* reject. Therefore, player 1's equilibrium beliefs about 2's type in $t = 1$, $\mu_1^{*1}(x_0)$, must be a uniform distribution on $[0, \theta^*]$. Using that $F_\theta^1(x|x_0) = \frac{x}{\theta^*}$ for $x \in [0, \theta^*]$, player 1's optimal strategy s_1^* must satisfy

$$s_1^{*1}(x_0) = \arg\max_{x \in X} x\left(1 - F_\theta^1(x|x_0)\right) = \frac{\theta^*}{2}.$$

This has to be substituted for x_1 in (1.28), since, in equilibrium, player 2's decision has to be optimal given x_0 and $s_1^{*1}(x_0)$.

---

[48] This, and also the following, is based on the tie-breaking assumption that player 2 accepts if he is indifferent. Since the event $\{\tilde{\theta} = x_1\}$ has zero probability, this does not affect players' expected payoffs.

Two cases have to be distinguished. First, assume $x_0 > (x_0 - \delta_2 x_1)/(1 - \delta_2)$. Then, $x_1 = \frac{\theta^*}{2}$ produces a contradiction to $\theta^* = x_0$. So, the second case, $\theta^* = (x_0 - \delta_2 x_1)/(1 - \delta_2)$, must be true. Using $x_1 = \frac{\theta^*}{2}$ yields $\theta^* = \frac{2x_0}{2 - \delta_2}$, implying

$$s_2^{*0}(x_0) = \begin{cases} 1; & \theta \geq \frac{2x_0}{2-\delta_2} \\ 0; & \theta < \frac{2x_0}{2-\delta_2} \end{cases}$$

and

$$s_1^{*1}(x_0) = \frac{x_0}{2 - \delta_2}.$$

It remains to determine s_1^{*0}. This must satisfy

$$s_1^{*0} = \underset{x \in X}{\arg\max} \, \underbrace{x \left(1 - F_\theta^0 \left(\frac{2x}{2-\delta_2}\right)\right)}_{\text{agreement in } t=0} + \underbrace{\frac{\delta_1 x}{2-\delta_2} \left(F_\theta^0\left(\frac{2x}{2-\delta_2}\right) - F_\theta^0\left(\frac{x}{2-\delta_2}\right)\right)}_{\text{agreement in } t=1}$$

which yields

$$s_1^{*0} = \frac{(2-\delta_2)^2}{8 - 4\delta_2 - 2\delta_1}.$$

Thus the essentially unique PBE of above two-stage bargaining game with incomplete information, the assessment $(s_1^*, s_2^*; \{\mu_1^{*0}, \mu_1^{*1}\}, \mu_2^*)$, is completely defined.[49] It is also the essentially unique sequential equilibrium (SE).

The equilibrium has several properties that are characteristic of many models of bargaining with incomplete information. For instance, the equilibrium involves *screening* of the informed player's type by the uninformed party: player 1 starts with a high price $s_1^{*0} \in X$. If player 2 has a high valuation of the good, i.e. $\theta \in [\theta^*, 1]$, he accepts immediately. Types below the threshold θ^* lose comparatively little from waiting, and prefer to a hold on for price $s_1^{*1}(s_1^{*0}) < s_1^{*0}$ in $t = 1$. So, player 1's strategy induces *inefficient delay*. This is individually rational because it yields a more favourable distribution of the (diminished) surplus. In models where the informed player makes more complex moves than just to accept or reject, delay can also occur as the result of *signalling*. Low valuation buyers may, for example, wait long before responding to offers in order to reveal that they have low costs of delay and will only buy at a low price.

The uncertainty with respect to player 2's preferences robs player 1 to a considerable extent of the last mover advantage that she would have under perfect information. Her take-it-or-leave-it offer in $t = 1$ gives a positive *information rent* to almost every informed player 2 who accepts in $t = 1$. This indicates that, on the one hand, a lack of information has a detrimental effect on one's bargaining power. On the other hand, player 1 could increase her expected surplus if she could commit to ignoring the information that is

---

[49] The qualification 'essentially' refers to the possibility to make different tie-breaking assumptions.

revealed by a rejection of her offer in $t = 0$.[50] Sobel and Takahashi (1983) investigate this detrimental effect of possessing information in detail.

Sobel and Takahashi (1983) also consider the extension of above model to more general type distributions and an infinite bargaining horizon. As in the two-stage model, player 1's price offers monotonically decrease over time; the bigger player 2's valuation and impatience, the earlier he trades. These two features are also referred to as *Coasian dynamics*. When $\delta_1 = \delta_2 = \delta$ and the length of a bargaining period, τ, approaches zero, the above-mentioned inefficiency due to delay vanishes. Moreover, the surplus from trade going to player 1 vanishes. Thus, the so-called *Coase conjecture* holds in Sobel and Takahashi's model: As transaction costs become negligible, bilateral negotiation results in an efficient outcome and the entire surplus is appropriated by the informed party.

In the two-stage model, a non-negligible fraction of types $\theta \in [0, \frac{2-\delta_2}{8-4\delta_2-2\delta_1})$ reject both equilibrium offers. So, given buyers with a low valuation, there is *inefficiency from too little trade* – possible mutual gains are not even realized with delay. For above distribution assumptions, this type of inefficiency is an artefact of the restriction to two stages. It does not occur for infinite bargaining horizon, in line with the Coase conjecture.

However, different assumptions on the distribution of buyer and seller valuations can invalidate the Coase conjecture. Myerson and Satterthwaite (1983) consider *individually rational and efficient mechanisms* without exogenous subsidies, which are the abstract representation of any possible efficient bilateral bargaining procedure. Myerson and Satterthwaite show that a positive probability of no gains from trade implies that no such mechanism exists. In realizations in which the buyer has a lower valuation than the seller, no-trade is efficient. However, the positive probability of no gains from trade implies that there is also no trade for a positive measure of buyers with a valuation actually exceeding the seller's reservation price. The reason is that any such mechanism[51] involves *incentive costs* incurred by the uninformed player in order to extract correct information. These costs put a lower bound on gains that can actually be realized by voluntary and un-subsidized trade.

The assumption that the informed player only makes binary choices facilitates above analysis. Typically, it helps to obtain a rather small set of equilibrium assessments: If in above example the informed player deviates from his equilibrium strategy and accepts an offer, then the game ends. If he deviates and rejects an offer, the uninformed player is unable to recognize the deviation because some types of the informed player would rationally have rejected; so beliefs are always updated by Bayes' rule.

In contrast, consider a different two-stage game and suppose that player 1 is an informed buyer, has a valuation $\theta \in \Theta = \{\theta_L, \theta_H\}$ with $\theta_L < \theta_H$, and

---

[50] This is, of course, excluded by the PBE or SE concepts.

[51] By the *revelation principle* (cf. e.g. Fudenberg and Tirole 1991, pp. 253ff), attention can be restricted to truth-revealing mechanisms.

makes the offer x_0 in period $t = 0$. Restricting attention to pure strategies, there are only two – possibly identical – offers, x_0^L and x_0^H, that the low- and high-valuation buyers, respectively, will make in $t = 0$ in equilibrium. Now suppose an offer $x_0 \in (x_0^L, x_0^H)$ is observed by the seller, player 2. His equilibrium strategy s_2^* must specify a continuation plan of actions which is optimal given his updated beliefs. However, the PBE concept imposes no restriction on updated beliefs for information sets which are reached with zero probability under the equilibrium strategies. These are also referred to as *off-equilibrium beliefs* or *conjectures*. For example, player 2 could have *optimistic conjectures* and infer that player 1 is of the high type θ_H. Then, unless player 2 is very impatient, it is sequentially rational to reject and to make the counter-offer $x_1 = \theta_H$. Thus, a threat to offer $x_1 = \theta_H$ in the final stage can be rationalized, supporting equilibria with very high offers $x_0^L = x_0^H$. Also, player 2 could have *pessimistic conjectures* and conclude from a deviation that player 1 is of the low type, and must therefore accept very low offers in $t = 0$. Another alternative would be the use of *passive conjectures*, meaning that player 2 keeps his old beliefs after a deviation.

This illustrates how the freedom in choosing off-equilibrium beliefs can serve to sustain multiple PBE, and also SE. The arbitrariness of bargaining predictions which has been overcome by imposing sequential rationality in the complete information case – considering SPE instead of NE – re-emerges in the case of incomplete information, because many different sets of off-equilibrium beliefs can be used to render many different strategies sequentially rational.

Rubinstein (1985b) investigates the role of conjectures in a modification of the infinite-horizon game of Sect. 1.3.2 in which player 1 has commonly known constant costs of delay c_1, and player 2 has privately known constant costs that are given by $c_2 = c_w > c_1$ with probability p, and equal $c_2 = c_s < c_1$ otherwise. A great variety of SE is defined by different types of conjectures. This indeterminacy can be overcome by imposing axioms that beliefs (and strategies) have to satisfy. Thus, Rubinstein (1985a) reduces the set of SE to a singleton. These axioms are plausible but not compelling, and Rubinstein later states (1987, p. 215):

> My intuition is that something is basically wrong in our approach to games with incomplete information and the "state of the art" of bargaining reflects our more general confusion.

1.4 Evolutionary Models

The predictions of both the cooperative and the non-cooperative approach to bilateral bargaining are based on a number of demanding assumptions about the two agents. For example, Rubinstein's (1982) SPE analysis supposes that

both parties fully understand the game and make correct, costless calculations of high complexity which result in consistent and sequentially rational
plans of actions for any eventuality. As a prerequisite, each player not only
needs to know his own complete and transitive preferences with respect to his
surplus share and possible agreement times, but also the preferences of his
opponent. Assuming imperfect information is not much of a relaxation since
a sequential equilibrium requires players to have common priors about the
distribution of surplus, time, and risk preferences, plus a coherent model of
the opponent's inference process. In practice, mutually optimal behaviour as
formalized by the NE concept – not to speak of its refinements – or the efficiency assumed by most cooperative bargaining solutions cannot be taken for
granted. The paradigm of a hyper-rational homo economicus produces several useful results, but are its predictions also relevant under less restrictive
assumptions about economic and social behaviour?

Evolutionary bargaining theory deals with this question. It is founded on
a different paradigm which assumes only *bounded rationality* a priori. The
approach focuses on feedback dynamics and the co-evolution of entire populations of agents, who adapt to a changing environment which they jointly create. Thus, *social norms—seedistribution norms* or *conventions* may emerge
which specify how a surplus is efficiently shared in frequently encountered
bargaining situations. Agents are assumed to have only partial understanding of the game they play. They use *rules of thumb* which are, at best, locally
optimizing and involve only a heuristic exploitation of available information.
In some models, agents are considered to occasionally act in completely arbitrary ways. These random events are referred to as *perturbations*. They
model trivial mistakes, reflect a possibly beneficial tendency to experiment,
or account for different types of un-modelled idiosyncracies.

Predictions of evolutionary models are based on the premise that the
endogenous adjustment process is fast relative to exogenous changes to the
system. Typically, a *recurrent game* is considered, meaning that a fixed non-
cooperative bargaining game is played infinitely often. Because players are
assumed to act myopically, the traditional repeated game effects are neglected.[52]

Agents' play is described in terms of states of a *dynamic system* whose
evolution over time is mathematically described. The system can be deterministic or stochastic. An example of the former is the *replicator dynamics*
(see the appendix). There, predictions are made in terms of rest points of difference or differential equations, their stability properties, and possibly the
speed at which distinguished states are approached or left. Stochastic evolutionary bargaining models are typically described by discrete-time stochastic
difference equations. One distinguishes the case of *fading noise*, in which the
system becomes essentially deterministic in the course of time, and models

---

[52] Compare e. g. Osborne and Rubinstein (1994, ch. 8) for Folk theorems for repeated games which are based on the paradigm of perfectly rational players.

with *persistent noise*. Predictions concern the average and long-run behaviour of the system. They are typically stated in terms of distinguished distributions over states, which may be approached dependent or independent of the initial state.

Agents' adaptation behaviour can take many forms. One example is the assumption of *myopic best replies*: Each agent behaves optimally given his present beliefs about others' behaviour, but does not anticipate strategic consequences of his behaviour as implied by the interdependence of agents' decision problems. This assumption is used in models of *fictitious play* (cf. e. g. Fudenberg and Levine 1998, chs. 2 and 4) and Young's (1993a) model of *adaptive play* introduced below. From the agent's point of view, this heuristic has the advantage of some robustness to exploitation. However, the (unmodelled) computational and informational costs may still be big. A smaller computational effort is required in models of *imitation learning*.[53] Agents are assumed to have groups of peers – possibly consisting of the entire population – whose behaviour is observed and evaluated. Successful members of the group then tend to be imitated by their less fortunate peers. Such behaviour is often the story underlying economists' use of replicator dynamics (see Sect. 1.4.2). Both myopic best replies and imitation can be combined with *experimentation*. The term refers to the random trial of available strategies irrespective of their present or past performance in the population. This is also a common feature of models of *reinforcement learning* in which each agent introspectively judges and chooses his strategies based on their past performance. Such behaviour has minimal computational and informational costs, but is vulnerable to exploitation by agents that act strategically. An example of reinforcement behaviour is investigated in Chap. 2.

Three evolutionary models of bargaining will now be considered. The first example is the stochastic evolution of bargaining conventions investigated by Young (1993b). His model establishes that the asymmetric Nash bargaining solution of Sect. 1.2.1 can serve as a good predictor of bargaining outcomes also under the assumption of boundedly rational agents. Then, an evolutionary analysis of a modification of Sect. 1.3.1's ultimatum game by Gale, Binmore, and Samuelson (1995) is sketched. Their replicator dynamics model illustrates the credentials of NE and SPE predictions under low rationality. Finally, the evolutionary foundation which Binmore, Piccione, and Samuelson (1998) propose for Rubinstein's (1982) SPE prediction of Sect. 1.3.2 is discussed.

1.4.1 Adaptive Play

Young's (1993b) path-breaking model uses the general framework of *adaptive play* in order to investigate bilateral bargaining. This framework – developed

---

[53] Compare Schlag (1998) for a survey, and a microeconomic justification of imitation.

in detail by Young (1993a) – assumes two finite, possibly overlapping classes or populations of agents, I_1 and I_2. In each period $t \in T = \{0, 1, 2, \ldots\}$, two randomly drawn agents, $j \in I_1$ and $k \in I_2$, play a 2-player finite game form $\langle \{1, 2\}, \Delta(S_1, S_2) \rangle$ in the role of player 1 and 2, respectively. Periods need not have identical length; they capture that interaction is sequential. In the basic setting, it is assumed that any agent in I_1 has positive probability of interacting with any agent in I_2 and vice versa; this can be relaxed to arbitrary interaction structures. Each agent $i \in I = I_1 \cup I_2$ has a von Neumann-Morgenstern utility function $\pi_i \colon \Delta(S_1, S_2) \to \mathbb{R}$ which is his private knowledge. The distributions of preferences in the populations are not known to any player a priori. So, beliefs about the opponent player have to be based on past observations. Young makes the assumption that no more than the $m \geq 1$ most recent observations are available. The collection $h_t = (s^{t-m+1}, \ldots, s^t)$ of the realized pure strategy combinations in the last m periods is called society's *memory* in period t. The memory is anonymous, i. e. strategy profiles are tracked but not agents.

Agents are assumed to exploit only part of the memory when making decisions. This can be motivated for example by agents' limited information processing capacities, or the assumption that agents receive information about past interaction only from a subgroup of past players, e. g. a sample of their friends, colleagues, or neighbours. So each agent $i \in I$ is characterized by a *sampling ratio* $r_i \leq 1$ which denotes the rational fraction of the available m observations that i processes. It is assumed that $r_i m$ is always an integer; values of m which satisfy this requirement will be referred to as *admissible memory sizes*. More restrictively, any memory sample of size $r_i m$ has positive probability to be drawn by i.

An agent $j \in I_1$ who is drawn to be player 1 in the game $\langle \{1, 2\}, \Delta(S_1, S_2), (\pi_j, \pi_k) \rangle$ played in period $t + 1$ uses the frequency distribution over S_2 in her *sample* $(s_2^{t_1}, \ldots, s_2^{t_{r_j m}})$ of player 2's strategy choices drawn from memory h_t as her beliefs about the choice of agent $k \in I_2$. Agent j then chooses $\sigma_1^{t+1} \in \Delta(S_1)$ to maximize her expected payoff given her beliefs. This is equivalent to playing a best reply to the mixed strategy $\sigma_2 \in \Delta(S_2)$ defined by her sample. Such behaviour is myopic because it neglects any implications of this choice for $t+2$, $t+3$, etc., but is hardly irrational if I_1 and I_2 are sufficiently large. It is an important assumption that σ_1^{t+1} always places positive probability on any pure strategy $s_1 \in S_1$ which is a best reply to the sample. Agent k acts analogously in the role of player 2. After j and k have chosen their mixed strategies, a pure strategy profile s^{t+1} is realized. It defines the new memory $h_{t+1} = (s^{t-m+2}, \ldots, s^t, s^{t+1})$, from which two agents j' and k' will draw their samples and calculate their choices.

Above dynamic process will be referred to as *unperturbed adaptive play*. Given some initial history $h_0 \in H = (S_1 \times S_2)^m$, it defines a discrete-time homogeneous *Markov process* Φ^0 on the finite state space H. The probabilities of transitions from memory $h \in H$ to memory $h' \in H$ can be summarized by a transition matrix $P = (p_{h\,h'})_{h,h' \in H}$.[54] It maps a distribution over states in period t, row vector $\mu_t \in \Delta(H)$, to the distribution $\mu_{t+1} \equiv \mu_t P$ over states in period $t + 1$. Iterating this difference equation, μ_t is also defined by a – possibly degenerated – initial distribution μ_0 over states and $\mu_t \equiv \mu_0 P^t$. Elements of P^t will be denoted by $p_{h\,h'}^{(t)}$. A state $h' \in H$ is called *accessible from* $h \in H$ if there is a positive probability that the process moves from h to h' in finitely many steps, i.e. there exists some t such that $p_{h\,h'}^{(t)} > 0$. The state space of a finite Markov process can always be partitioned into $n \geq 1$ *recurrence classes* $C_1, \ldots, C_n$, each of which has the property that from any state $h \in C_i$ any state $h' \in C_i$ is accessible but no state $h'' \notin C_i$ is accessible, and one class $\bar{C}$ of transient states.

If a process has only one recurrent class $C \subseteq H$[55] and is *aperiodic*, i.e. $(\exists h \in C)(\exists t_0 \in T)(\forall t \geq t_0) : p_{h\,h}^{(t)} > 0$, whereby quasi-deterministic cycles through C are excluded, then there exists exactly one *stationary distribution* $\mu' \in \Delta(H)$ for which $\mu' = \mu' P$. Moreover, first, the empirical frequency distribution over H of (almost) every realization of the process and, second, the theoretical t-step distribution $\mu_0 P^t$ both converge to μ' for $t \to \infty$ independent of the initial distribution μ_0. A stochastic process with these two convergence properties is called *ergodic*. Ergodicity implies that a very precise prediction of the long-run dynamics of the process can be given.

Unfortunately, unperturbed adaptive play typically has multiple recurrence classes. In particular, every memory which consists of m repetitions of a strict Nash equilibrium $s^* \in S_1 \times S_2$ is a recurrence class. For games with multiple strict NE, this implies that long-run dynamics crucially depend on the initial state $h_0 \in H$.

This is no longer the case if agents are assumed to occasionally choose any strategy at random, i.e. if the process is perturbed. Suppose, for example, that each player $i \in I$ independently experiences a perturbation with a probability $\lambda_i \eta > 0$, and conditional on that event he chooses each element of his respective strategy set with positive probability.[56] H itself is the unique recurrence class of this perturbed Markov process Φ^η – referred to as *(perturbed) adaptive play*. Φ^η is ergodic, and a unique stationary distribution μ^η describes its long-run evolution. So, the realistic feature that agents make mistakes and occasionally behave unpredictably has a useful technical implication.

Different *noise levels* η may be realistic in different models. In any case, the strategy profiles forming states $h \in H$ which have high probability $\mu^\eta(h)$ for small noise levels are distinguished as particularly stable and prominent choice combinations under the assumed adaptation behaviour. If

---

[54] Good introductions to Markov processes can be found in Taylor and Karlin (1998) and Stokey and Lucas (1989, chs. 8, 11–12).

[55] This is a weaker requirement than *irreducibility*.

[56] More generally, it suffices to assume a big enough support of perturbed choices such that any $h' \in H$ becomes accessible from arbitrary $h \in H$.

$\lim_{\eta \to 0} \mu^{\eta} = \mu^*$ exists, then μ^* places weight on a subset of states belonging to the recurrence classes of Φ^0. States $h \in H$ with $\mu^*(h) > 0$ have been named *stochastically stable states (SSS)* by Foster and Young (1990). If each recurrence class of Φ^0 is a memory consisting of the m-fold repetition of a strict Nash equilibrium, then SSS of Φ^{η} are a powerful criterion of *equilibrium selection* which is based on the assumption not of more rationality – as are SPE, SE, etc. – but less. In case of a unique SSS, this state is observed in (almost) every process realization with a time share close to one.

The identification of SSS is non-trivial. A powerful corresponding analytical framework is developed by Young (1993a) – combining results on the characterization of stationary distributions by Freidlin and Wentzell (1984) and graph-theoretic arguments. A good exposition can be found in Young (1998, ch. 3). The basic idea is to count the minimal number of perturbations that are needed to move from states in one recurrence class C_k directly or indirectly to states in any other recurrence class C_l of the unperturbed process. This minimal number of perturbations is also called the *resistance* r_{kl} of a transition from C_k to C_l. For any fixed recurrence class C_i, the resistances are then aggregated to the *stochastic potential* of C_i, which, loosely stated, summarizes the least resistant way to reach C_i from all other recurrence classes.[57] Young (1993a) establishes that the SSS of adaptive play Φ^{η} are precisely the states contained in the recurrence classes of unperturbed adaptive play Φ^0 which have minimal stochastic potential.

This framework shall now be applied to bilateral bargaining. The presentation is based on Young (1993b) and also Young (1998, ch. 8). Assume that the finite populations, I_1 and I_2, are disjoint. For example, let them correspond to landlords and tenant farmers, respectively, who bargain about shares of a crop that is grown and harvested in period t in case of agreement. Let each population be perfectly homogeneous, i.e. $r_j = r_1$, $\pi_j = \pi_1$, and $\lambda_j = 1$ for every landlord $j \in I_1$, and analogously $r_k = r_2$, $\pi_k = \pi_2$, and $\lambda_k = 1$ for every tenant farmer $k \in I_2$, where π_1 and π_2 are weakly concave and strictly increasing in the agent's share. The assumption of homogeneous noise does not affect results; the case of heterogeneous preferences and sampling ratios will be indicated later.

In each period t, the drawn landlord and tenant farmer play a discrete version of the *demand game* proposed by Nash (1953). Both agents simultaneously demand shares $x_t, y_t \in X_{\xi} = \{\xi, 2\xi, \ldots, 1\}$, respectively. It will be assumed that $1/\xi$ is an integer. If agents' demands are compatible, i.e. $x_t + y_t \leq 1$, the payoffs $\pi_1(x_t)$ and $\pi_2(y_t)$ result. If the demands are incompatible, both players receive $\pi_1(0) = \pi_2(0) = 0$. The unperturbed and perturbed processes of adaptive play defined by these assumptions will be referred to as

---

[57] More precisely, a *tree rooted at* C_i is a directed graph which has Φ^0's recurrence classes as nodes, contains a unique path from any $C_j \neq C_i$ to C_i, and has resistance r_{kl} as the weight of the edge between any C_k and C_l. The minimal sum of edge weights of all trees rooted at C_i is C_i's stochastic potential.

the *unperturbed bargaining process* Φ^0 and the *perturbed bargaining process* Φ^η, respectively.

The strategy profiles $(\xi, 1 - \xi), \ldots, (1 - \xi, \xi)$ are the strict NE of the demand game. Memories $c_x = ((x, 1 - x), \ldots, (x, 1 - x)) \in H$ that consist of the m-fold repetition of a strict NE will be referred to as a *convention*. Clearly, every convention c_x for $x \in \{\xi, \ldots, 1 - \xi\}$ is an *absorbing state* of the unperturbed bargaining process Φ^0, i.e. c_x will never be left if it is reached. Hence, every set $\{c_x\}$ constitutes a singleton recurrence class. In fact, these are the only recurrence classes of Φ^0 if $r_1, r_2 \leq \frac{1}{2}$.[58] The way to see this is to consider an arbitrary state $h_t \in H$ which is not a convention. Then, a finite chain of samples and corresponding successor states to h_t can be constructed such that each sample and each transition has positive probability, and the last transition is to a convention. It follows from standard results for finite Markov processes that Φ^0 converges to some convention c_x with probability one.

The probability of converging to a particular convention c_x depends on the initial state h_0, and is trivially unity if $h_0 = c_x$. So, any crop-sharing convention can emerge in Φ^0. However, a sharp long-run prediction for the bargaining result is obtained for the perturbed bargaining process Φ^η:

Theorem 1.4. (Young) *For any smallest unit $\xi > 0$ of demands such that $1/\xi$ is an integer, the perturbed bargaining process Φ^η defined for disjoint homogeneous populations with sampling ratios $r_1, r_2 \leq \frac{1}{2}$ has at least one and at most two stochastically stable states for all sufficiently large admissible memory sizes m. These SSS are conventions, and for $\xi \to 0$ the underlying NE payoffs $(x, 1 - x)$ converge to the utility levels of the asymmetric Nash bargaining solution $F^{N(\alpha, \beta)}$ with bargaining powers $\alpha = r_1$ and $\beta = r_2$[59] of the bargaining problem $\langle U(\pi_1, \pi_2), 0 \rangle$ defined by*

$$U(\pi_1, \pi_2) = \left\{ (u_1, u_2) \in \mathbb{R}^2 : u_1 \leq \pi_1(x) \wedge u_2 \leq \pi_2(1 - x) \wedge x \in [0, 1] \right\}.$$

Sketch of the proof: The reader is referred to Young (1993b) for a rigorous proof of this result. As indicated above, the fundamental determinant of the robustness of a convention c_x is the minimal number of perturbations in Φ^η needed to switch from c_x to some other convention $c_{x'}$. Such a switch can result from four types of 'mistake' accumulations caused by perturbations. A first possibility is that landlords make too low demands $x' < x$ for $f_1(x, x') \in \mathbb{N}$ times in a row, and that it then becomes optimal for a tenant farmer who has drawn an appropriate sample to increase his demand from $1 - x$ to $1 - x'$. Given the restriction on sampling ratios r_1 and r_2, convention $c_{x'}$ is then accessible without another perturbation. The relevant sample consists of $f_1(x, x')$ times the demand x' and $mr_2 - f_1(x, x')$ times the demand

---

[58] For heterogeneous agents, it suffices that at least one agent in each population samples no more than half of the memory.

[59] See p. 16 for the general definition of $F^{N(\alpha, \beta)}$ which does not require $\alpha + \beta = 1$.

x. Player 2's subjectively expected payoff from repeating his conventional demand $1-x$ is $\pi_2(1-x)$ because it will certainly produce compatible demands. The demand $1-x'$ yields expected payoff $\frac{f_1(x,x')}{mr_2}\pi_2(1-x')$ according to 2's beliefs. So, player 2's best response is $1-x'$ if and only if

$$f_1(x,x') \geq m\,r_2 \frac{\pi_2(1-x)}{\pi_2(1-x')}.$$

Of all possible mistaken demands $x' < x$ by the landlords, $x' = \xi$ minimizes the right-hand-side. Hence, for any convention c_x the term $\lceil f_1(x)\rceil$ with

$$f_1(x) = m\,r_2 \frac{\pi_2(1-x)}{\pi_2(1-\xi)},$$

is the minimal number of perturbations required to switch from c_x to some other convention by this type of mistake.[60]

   Similar calculations can be carried out for the other types of mistakes: Let $f_2(x,x')$ denote the number of perturbations that are needed to switch from c_x to $c_{x'}$ in case that tenants successively make aggressive demands $1-x' > 1-x$. Calculating the minimum of $f_2(x,x')$ for $x' \in \{\xi,\ldots,x-\xi\}$ yields $\lceil f_2(x)\rceil$ with

$$f_2(x) = m\,r_1\left(1 - \frac{\pi_1(x-\xi)}{\pi_1(x)}\right) \tag{1.29}$$

as the minimal number of perturbations required by this way of switching away from c_x. Expression $f_2(x)$ is precisely the landlords' sample size times their relative loss of utility from giving up a ξ-increment of their share in c_x. Similarly, the minimal resistance to switch from c_x to some convention that is less favourable to the tenants in consequence of aggressive demands by the landlords can be calculated. It is at least the tenants' sample size times their relative loss of utility from giving up a ξ-increment, i.e. $\lceil f_3(x)\rceil$ with

$$f_3(x) = m\,r_2\left(1 - \frac{\pi_2(1-x-\xi)}{\pi_2(1-x)}\right). \tag{1.30}$$

The remaining possibility is that tenants by mistake make too low demands $1-x' < 1-x$ several times in a row. This takes at minimum $\lceil f_4(x)\rceil$ perturbations with

$$f_4(x) = m\,r_1 \frac{\pi_1(x)}{\pi_1(1-\xi)}.$$

Allowing for all types of mistake accumulations, it turns out to require a minimal amount of $\lceil r^\xi(x)\rceil$ perturbations with

---

[60] $\lceil y\rceil$ denotes the smallest integer greater than or equal to y. The strict monotonicity of π_2 and the requirement that m is sufficiently large ensure that $\lceil f_1(\cdot)\rceil$, etc. are one-to-one.

$$r^\xi(x) = \min\{f_1(x), f_2(x), f_3(x), f_4(x)\} \tag{1.31}$$

to switch from convention c_x to some other convention. Closer investigation turns out that $f_4(x)$ is never strictly smaller than $f_2(x)$, and so can be dropped from (1.31).

Young (1993b) shows that those conventions c_x for which $\lceil r^\xi(x)\rceil$ is maximal have minimal stochastic potential. Hence, they are the stochastically stable states. Using the monotonicity and concavity properties of π_1 and π_2, $r^\xi(x)$ can be shown to be unimodal on $[0,1]$, i.e. there is a unique maximizer $x^{\xi*}$ on $[0,1]$ and at most two maximizers of $r^\xi(x)$ or $\lceil r^\xi(x)\rceil$ on $\{\xi, 2\xi, \ldots, 1-\xi\}$ given sufficiently large memory size m. The latter are arbitrarily close to $x^{\xi*}$ if ξ is small.

Now consider,

$$f^\xi(x) = r^\xi(x)/\xi = \min\{f_1(x)/\xi, f_2(x)/\xi, f_3(x)/\xi\}. \tag{1.32}$$

Since r^ξ and f^ξ differ only by the constant factor $1/\xi$, they have the same maximizer $x^{\xi*}$ on $[0,1]$ for any $\xi > 0$. The term $f_1(x)/\xi$ goes to infinity for $\xi \to 0$. For sufficiently small $\xi > 0$ it can be dropped in (1.32), and $f^\xi(x)$ is maximized at the intersection of decreasing function $f_2(x)/\xi$ and increasing function $f_3(x)/\xi$. Thus $f_2(x^{\xi*})/\xi = f_3(x^{\xi*})/\xi$ holds. Assuming that π_1 and π_2 are continuously differentiable,[61] one gets

$$\lim_{\xi\to 0} f_2(x)/\xi = m\,r_1 \frac{\partial\pi_1(x)/\partial x}{\pi_1(x)}$$

and

$$\lim_{\xi\to 0} f_3(x)/\xi = m\,r_2 \frac{\partial\pi_2(1-x)/\partial x}{\pi_2(1-x)}.$$

Since $x^{\xi*}$ satisfies $f_2(x)/\xi = f_3(x)/\xi$ for any sufficiently small ξ, its limit, $x^* \in (0,1)$, satisfies

$$r_1 \frac{\partial\pi_1(x^*)/\partial x}{\pi_1(x)} = r_2 \frac{\partial\pi_2(1-x^*)/\partial x}{\pi_2(1-x)}.$$

This is the necessary and sufficient condition for x^* to maximize

$$r_1 \ln\pi_1(x) + r_2 \ln\pi_2(1-x),$$

i.e. for $(x^*, 1-x^*)$ to be the asymmetric Nash solution with bargaining powers r_1 and r_2 of above bargaining problem. □

The theorem establishes that if the unit in which demands can be made is sufficiently small and society has long memory, then average payoffs from playing perturbed myopic best replies are arbitrarily close to

---

[61] Young (1993b) shows that subdifferentiability is, in fact, sufficient.

$F^{N(r_1, r_2)}(U(\pi_1, \pi_2), 0)$. Moreover, recalling the definition of a SSS, a surplus division approximating the asymmetric Nash solution will actually be observed as the society's surplus-sharing convention almost all of the time. It deserves emphasis that no mutual or even common knowledge of players' preferences is assumed. Rather, players' adaptive behaviour based on privately known preferences and sampling ratios gives rise to this particular surplus division.

Equations (1.29) and (1.30) highlight the driving force behind this result – namely, players' individual resistance to giving up a ξ-increment of surplus in view of samples with aggressive demands by the opponent. The more perturbations resulting in sampled aggressive demands are necessary to upset convention c_x, the more stable it is.[62] For given c_x, the minimum length of a successful run of spontaneously increased demands is the opponents' sample size times his relative loss of utility from giving up a ξ-increment. A balance of power is reached for the division(s) $(x, 1-x)$ for which this weighted relative loss is equal for both players.

It follows that, ceteris paribus, the stable surplus-sharing convention favours a population that exploits more information – corresponding to greater sampling ratio r_i – and that is less risk averse: The behaviour of a player who draws a larger sample is less easily influenced by aggressive mistakes by the other side. And the less risk averse a player is, the greater is the number of aggressive mistakes by the opponent before the player stops betting on the more favourable established convention.

Nothing has been assumed about the sizes of populations, except that they are finite. So, Theorem 1.4 also captures the case of only two agents who repeatedly bargain about a renewable surplus. The assumption concerning the partial sampling of information is then clearly less natural than for two large populations. Still, it is noteworthy that though the respective settings, information and rationality assumptions, and the ways of reasoning are completely unrelated, the asymmetric Nash solution is corroborated as a reasonable predictor of the result of bilateral bargaining once more.

Theorem 1.4 remains true when unclaimed surplus in the demand game is split equally between the players. This removes the demand game's unnatural feature of leaving surplus on the table. The perfect homogeneity of the two populations is not very realistic either. A more general setting is to allow players to have individually distinct preferences – represented by weakly concave, non-negative, and strictly increasing von Neumann-Morgenstern utility functions – and distinct sampling ratios. This can be modelled by identifying each agent $i \in I$ with a pair (r_i, π_i). Young (1993b) shows that under assumptions analogous to Theorem 1.4, the unique stochastically stable division(s)

---

[62] One can infer that in case of asymmetric noise levels η_1 and η_2 in both populations, it is ceteris paribus beneficial e. g. for population 1 to have higher η_1. This questions the relevance of limit consideration $\eta \to 0$ if one imagines an encompassing evolutionary selection process as the determinant of agents' noise parameters.

converge to the unique maximum of the strictly quasi-concave function[63]

$$r(x) = \min\left\{\min_{j \in I_1} m\,r_j \frac{\partial \pi_j(x)/\partial x}{\pi_j(x)}, \min_{k \in I_2} m\,r_k \frac{\partial \pi_k(1-x)/\partial x}{\pi_k(1-x)}\right\}. \quad (1.33)$$

Again, weighted relative utility losses from giving up a ξ-increment are the driving force behind the result. The *Young bargaining solution* determined by (1.33) no longer corresponds to the asymmetric Nash solution and neither the related *generalized Nash solution*. The latter is introduced by Harsanyi and Selten (1972) to account for incomplete information about bargainers. It is based on the relative frequencies of player types. The Young solution, in contrast, depends only on the support of the type distribution. Loosely speaking, the 'weakest' individual in each population defines its long-run bargaining power.[64]

So far, it has been assumed that the populations of bargainers, I_1 and I_2, are disjoint. It seems reasonable to allow for some social mobility between classes, so that some agent j may originally be a tenant farmer and bargains as player 2, but later in the process becomes a landlord and bargains as player 1. If mixing between the classes creates a common support of type distributions – not necessarily the same distributions – for players 1 and 2, then (1.33) becomes a re-scaling of

$$\hat{r}(x) = \min_{i \in I}\left\{\frac{\partial \pi_i(x)/\partial x}{\pi_i(x)}, \frac{\partial \pi_i(1-x)/\partial x}{\pi_i(1-x)}\right\}.$$

This is maximized for $x = \frac{1}{2}$. So, the convention of approximately dividing a surplus fifty-fifty will be adhered to with very high probability provided that society's memory is large, the noise level is small, and there is some social mobility. The fifty-fifty division is clearly a *focal point* of bargaining in practice. This need not be grounded on any inherent fairness properties, explicitly symmetric bargaining powers, or an intrinsic "power to communicate its own inevitability to the two parties in such fashion that each appreciates that they both appreciate it" (Schelling 1960, p. 72). Rather, 'fifty-fifty' may simply be the most stable convention for dividing a jointly created surplus as in Young's model.

Sáez-Martí and Weibull (1999) investigate the robustness of Theorem 1.4 to the introduction of a share $\lambda > 0$ of 'clever' agents to population I_1. Clever agents are assumed to know utility function π_2. They try to anticipate the behaviour of player 2 by, first, sampling recent demands from their own population and, second, by playing a best reply to player 2's best reply to this sample. It turns out that a share $\lambda < 1$ of clever agents does not affect the result if $r_1 \geq r_2$. Only if $r_1 < r_2$ does population I_1 benefit from the

---

[63] Again, differentiability of utility functions is not essential.
[64] Clearly, the minima in (1.33) may generally be located at different agents $j \in I_1$ and $k \in I_2$ for different x.

cleverness of some of their members. Namely, irrespective of the size of $\lambda < 1$, the cleverness has the same effect as decreasing the sampling ratio r_2 to the lower level r_1, i.e. the symmetric Nash bargaining solution is approximated. In case that the entire population I_1 acts cleverly, $\lambda = 1$, but still experiences rare perturbations, it gets approximately the **whole surplus**.

The demand game has been introduced by Nash (1953) precisely in order to provide a non-cooperative foundation for his cooperative bargaining solution.[65] So it may be regarded as not surprising that Young (1993b) further corroborates the Nash solution based on this particular bargaining game. In fact, Young (1998, ch. 9) also provides some corroboration for the Kalai-Smorodinsky bargaining solution introduced in Sect. 1.2.2.

Namely, one can derive a *pure coordination demand game* from the Nash demand game in which each player $i = 1, 2$ proposes a possibly inefficient division (x_i, y_i) with $x_i, y_i \in \{\xi, 2\xi, \ldots, 1\}$ and $x_i + y_i \leq 1$. Agents receive the payoffs $\pi_1(x)$ and $\pi_2(y)$, respectively, if their proposals have been perfectly coordinated, i.e. if $(x_1, y_1) = (x_2, y_2) = (x, y)$, and zero otherwise. The requirement of perfect coordination introduces many inefficient conventions which affect the resistances between efficient conventions.[66] Allowing for global perturbations as before, the payoffs of the stochastically stable conventions in this game approximate the Kalai-Smorodinsky solution of the corresponding bargaining problem as $\xi \to 0$ (see Young 1998, pp. 141ff, for a more precise statement).

It questionable whether the pure coordination demand game is a particularly plausible model of bargaining, and hence whether this result should be interpreted as a meaningful corroboration of the Kalai-Smorodinsky solution. Young presents the pure coordination demand game as an illustration of the emergence of social contracts (see Chap. 4). Such contracts generally cover more than just the division of a surplus, and it may possibly make sense to assume that only perfect coordination yields a positive payoff. Young (1998, p. 143) summarizes adaptive play's tendency towards the Kalai-Smorodinsky contract, which is averaging or balancing both players' ideal contract:

> Change ... is driven by those who have the most to gain from change. Over the long run, this tends to favor contracts that are efficient and that offer each side fairly high payoffs within the set of payoffs that are possible.

1.4.2 Replicator Dynamics and the Ultimatum Minigame

The *ultimatum minigame* (UMG) is proposed by Gale, Binmore, and Samuelson (1995) as a more tractable, simplified version of the ultimatum game of

---

[65] Nash considers the set of NE of a perturbed version of the game in which the exact location of Pareto boundary $P(U)$ is uncertain. As the uncertainty is decreased and the perturbed game approaches the demand game, the set of NE converges to the symmetric Nash bargaining solution (cf. Binmore 1987a).

[66] Note that, for example, the profile $((0.2, 0.3), (0.2, 0.3))$ is a strict NE in the pure coordination game.

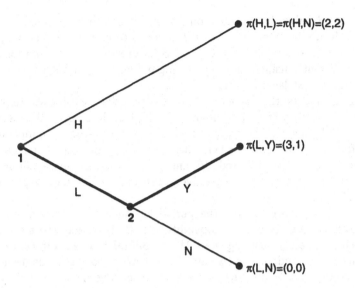

Fig. 1.13. The ultimatum minigame (UMG)

Fig. 1.8 (a) (p. 27). Player 1 moves first and can either propose a symmetric surplus division, which is automatically accepted by player 2 and yields the payoff vector $(2,2)$. Or she can propose an asymmetric surplus division, which yields the payoff combination $(3,1)$ if player 2 accepts this proposal, and $(0,0)$ otherwise. Players' pure strategy sets are $S_1 = \{H, L\}$ and $S_2 = \{Y, N\}$, respectively, where H denotes player 1 making the high offer and L the low offer. Similarly, Y indicates that both high and low offers are accepted by player 2, and N corresponds to the rejection of a low offer. The game is summarized in Fig. 1.13.

The UMG has two pure-strategy NE, (L, Y) and (H, N). The former equilibrium is strict and the unique SPE. The UMG also has a component of non-degenerated mixed-strategy NE. In each of them, player 1 chooses H with certainty, and player 2 accepts a low offer with some probability $p \in (0, \frac{2}{3}]$. The SPE (L, Y) is selected by iterative elimination of weakly dominated strategies, too. It is game theory's predicted outcome under the assumption of a perfectly informed player 1 who knows about player 2's rationality.

Gale et al. (1995) investigate the robustness of this prediction in an evolutionary setting that assumes two disjoint infinite populations, $I_1 = [0, 1]$ and $I_2 = [0, 1]$, of agents who repeatedly play the UMG in random pairs. Each agent acts like a simple stimulus-response mechanism with two modes of operation. In the main *playing mode*, an agent $k \in I_i$ simply carries out the strategy $s_k \in S_i$ whenever he receives the stimulus to play. The strategy s_k is taken as given in the playing mode, but occasionally revised in the *learning mode*. A revision means that agent k compares his average payoff from playing s_k with some exogenous, random *aspiration level* l_k. If his average payoff is lower than l_k, then he imitates some other agent l in his population. That can result in a reinstatement of s_k if l happens to use strategy $s_l = s_k$. The probability of choosing the strategy $s \in S_i$ is precisely the population share of agents presently using s in their playing mode. If agent k's average payoff from using s_k is at least l_k, then he sticks to it.

Let $x_t \in [0, 1]$ be the share of proposer population I_1 using strategy H in period $t \in T = \{0, \tau, 2\tau, \ldots\}$. Similarly, $y_t \in [0, 1]$ is the share of the responder population playing Y. The *state space* of this discrete-time dynamic system is therefore the unit square, where the respective proportions of L- and N-players can be directly inferred. H- and L-players' average payoffs in state $(x, y) \in [0, 1]^2$ are 2 and $3y$, respectively, and those of Y- and N-players are $1 + x$ and $2x$.

Gale et al. (1995) suppose the period length τ to be scaled such that each player has precisely the probability $\tau \in (0, 1]$ to get into his learning mode before he starts playing the UMG in period t. The aspiration level l_k is assumed to be independently uniformly distributed on the same interval $[\underline{l}_t, \overline{l}_t]$ with $0 \leq \underline{l}_t < 2$, $3y < \overline{l}_t \leq 3$ for each proposing agent $k \in I_1$ in period t.[67] Let $F^t_{l_1}$ be the common cumulative distribution function in period t. The resulting evolution of H's population share is described by the difference equation

$$x_{t+\tau} = \overbrace{x_t(1 - \tau)}^{\text{H-players not learning}} + x_t\tau \left[\overbrace{F^t_{l_1}(2)}^{\text{H-players satisfied}} + \overbrace{\left(1 - F^t_{l_1}(2)\right) x_t}^{\text{H-players imitating H}} \right.$$
$$\left. + \underbrace{(1 - x_t)\tau \left(1 - F^t_{l_1}(3y_t)\right) x_t}_{\text{L-players imitating H}} \right]$$

(1.34)

Note that the randomness of learning and aspirations plays no role since the share is almost surely the expected share given the infinite population size. (1.34) can be rearranged to

$$\frac{x_{t+\tau} - x_t}{\tau} = x_t \frac{2 - (2x_t + 3(1 - x_t)y_t)}{\overline{l}_t - \underline{l}_t}.$$

(1.35)

The corresponding equation for the responder population I_2 is

$$\frac{y_{t+\tau} - y_t}{\tau} = y_t \frac{(1 + x_t) - (2x_t + y_t - x_t y_t)}{\overline{l}'_t - \underline{l}'_t},$$

(1.36)

where all proposers' aspiration levels are independently distributed uniformly on the same interval $[\underline{l}'_t, \overline{l}'_t]$ with $0 \leq \underline{l}'_t < 1+x$, $2x < \overline{l}'_t \leq 2$. When one assumes that $\overline{l}_t - \underline{l}_t = \overline{l}'_t - \underline{l}'_t = 1$, difference equations (1.35) and (1.36) correspond to the continuous-time dynamic system

---

[67] Gale, Binmore, and Samuelson (1995) do not explicitly allow for time dependence of the aspiration distribution.

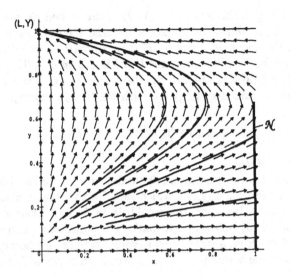

Fig. 1.14. Illustration of the UMG replicator dynamics (1.37)

$$\begin{pmatrix} \dot{x} \\ \dot{y} \end{pmatrix} = F(x,y) = \begin{pmatrix} [2 - (2x + 3y(1 - x))]x \\ [(1 + x) - (2x + y - xy)]y \end{pmatrix} \qquad (1.37)$$

where x abbreviates $x(t)$, and $\dot{x}$ is its first derivative with respect to t. The system of differential equations (1.37) is a two-population version of the *replicator dynamics* (see the appendix) of Taylor and Jonker (1978): The growth rate $\dot{x}/x$ of the share of H-players in population I_1 is precisely the positive or negative absolute difference between their expected payoff and the average payoff in their population.[68] The same holds for the share of Y-players in I_2.

Admittedly, above behavioural assumptions about agents are tailor-made to yield the replicator dynamics (1.37).[69] Often enough, economists use these well-investigated dynamics – originally developed by theoretical biologists – without being bothered to give a behavioural foundation at all. The quite specific model of exogenous aspiration levels above illustrates that replicator dynamics are meaningful also in non-biological contexts.

---

[68] An alternative two-population version of the replicator dynamics would equate $\dot{x}/x$ with the relative difference between H's expected payoff and the average of the population. Compare e. g. Weibull (1995, sec. 5.2) for details.

[69] These derivations are omitted in Samuelson (1997, ch. 5), which is a slightly varied exposition of Gale et al. (1995). Samuelson (1997, ch. 3) provides a motivation of replicator dynamics as the approximation of a stochastic process involving fixed aspiration levels but random payoffs. Benaim and Weibull (2000) give a number of powerful results about the approximation of several types of stochastic processes, which model boundedly-rational economic behaviour, by different versions of the replicator dynamics.

The dynamic process defined by (1.37) is illustrated for different initial conditions by the combined vector plot and phase diagram in Fig. 1.14.[70] The eigenvalues of the Jacobian matrix of F at $(x, y) = (0, 1)$ are both -1, and so the subgame perfect equilibrium of the UMG is *asymptotically stable* (cf. e. g. Simon and Blume 1994, p. 687). It can also be verified that the pure-strategy NE (H, N) and all mixed-strategy NE with $p \in (0, \frac{2}{3})$ – corresponding to the set

$$\mathcal{N} = \left\{ (x, y) \in [0, 1]^2 : x = 1 \wedge y < \frac{2}{3} \right\}$$

– are *Lyapunov stable.*

So the purely heuristic imitation dynamics turn out to attribute the highest stability to the profile (L, Y) which satisfies the most demanding rationality requirements. However, the equilibria in $\mathcal{N}$ have their stability, too. Even if (un-modelled) isolated perturbations or mutations – corresponding to spontaneous shifts from a state (x, y) to some state (x', y') nearby – occur, it can take a long time before the system moves from states in $\mathcal{N}$ to somewhere near the SPE.

The dynamic interpretation of asymptotic and Lyapunov stability refers to different degrees of robustness in response to isolated perturbations. It is worthwhile to investigate what happens in the presence of a small but persistent noise level, also called *drift*. Assume, for example, that agents in population I_1 act as specified above with probability $1 - \eta_1 \in (0, 1)$. With probability η_1, an agent $k \in I_1$ simply flips a coin and chooses H or L with equal probability. Making a similar assumption for the responder population, this modification leads to the following *perturbed replicator dynamics*:

$$\begin{pmatrix} \dot{x} \\ \dot{y} \end{pmatrix} = F^\eta(x, y) = \begin{pmatrix} (1 - \eta_1)[2 - (2x + 3(1 - x)y)]x \\ (1 - \eta_2)[(1 + x) - (2x + y - xy)]y \end{pmatrix} + \begin{pmatrix} \eta_1(\frac{1}{2} - x) \\ \eta_2(\frac{1}{2} - y) \end{pmatrix}$$
(1.38)

The dynamic system (1.38) is illustrated in Fig. 1.15 for the case in which proposer and responder population have an equal noise level of $\eta_1 = \eta_2 = 0.01$. It can be seen – and analytically confirmed – that states in $\mathcal{N}$ lose their stability through the introduction of the noise. This corroborates the SPE prediction for the ultimatum minigame. But as can be seen in Fig. 1.15, trajectories from a large subset of $[0, 1]^2$ first move towards the unstable component $\mathcal{N}$. It can be checked that the velocity of change (not indicated in Fig. 1.15) on trajectories approaching $\mathcal{N}$ is decreasing and gets close to zero before the turn towards $(0, 1)$. So it takes a long time before the system moves, for example, from $(0.15, 0.15)$ to the SPE $(0, 1)$.

Gale et al. (1995) distinguish four time spans in which different statements about the evolution of UMG strategy profiles apply. In the *short run*, the average surplus division between proposers and responders is entirely

---

[70] Figures 1.14–1.16 have been produced using the **DEtools** package of **Maple V R5**.

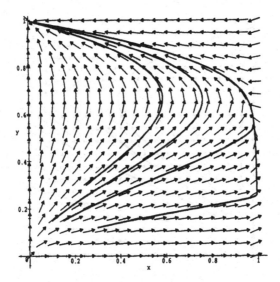

Fig. 1.15. Perturbed UMG replicator dynamics (1.38) with symmetric noise $\eta_1 = \eta_2 = 0.01$

defined by initial conditions, possibly corresponding to so-called *framing effects* observed in many laboratory experiments.[71] Then, in the *medium run* agents begin to learn and adapt their behaviour. This can – but generally need not – lead to convergence to an equilibrium of the game in the *long run*. In what Gale et al. (1995) call the *ultralong run*, perturbations would occasionally accumulate in a genuinely stochastic model, and possibly cause switches between different equilibria. This could select stochastically stable states (SSS).

As seen, the medium run may turn out be quite long for certain initial conditions. The corroboration of the SPE by the perturbed dynamics (1.38) is further weakened by the consideration of asymmetric noise levels in the populations. Figure 1.16 illustrates the case of $\eta_1 = 0.005$ and $\eta_2 = 0.05$, where a state close to $(1, 0.54)$ is also asymptotically stable. This provides a corroboration of symmetric proposals, and also sequentially irrational rejections of asymmetric offers. Gale et al. (1995) explicitly quantify the degrees of asymmetry between noise levels η_1 and η_2 that produce this result even as $\eta_1, \eta_2 \to 0$.

---

[71] Kahnemann and Tversky (1979), for example, investigate the significant differences in people's decisions when the same lotteries and net wealths are described or 'framed' with a different reference point – triggering different 'initial conditions' of subjects' behaviour through implicit and possibly wrong analogies. Camerer, Johnson, Rymon, and Sen (1993) investigate framing effects in sequential bargaining.

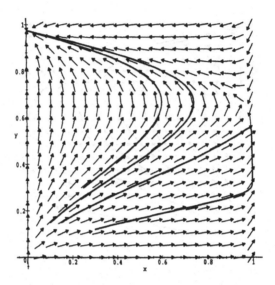

Fig. 1.16. Perturbed UMG replicator dynamics (1.38) asymmetric noise $\eta_1 = 0.005$ and $\eta_2 = 0.05$

Why should the noise level in one population be higher than in the other? In a further modification of the basic replicator equations (1.37), Gale et al. (1995) endogenize noise levels. In particular, η_i, $i = 1, 2$, is assumed to be a decreasing function of agents' *involvement* in playing the UMG. This involvement is then considered to be proportional to the absolute difference between the expected payoffs from playing H or L for population I_1, and from playing Y or N for population I_2. Since the expected payoff effect of a switch between Y and N is always at most 1, whereas that of H and L is at least 1 and ranges up to 3, proposers are generally more involved than responders according to this specification. It turns out that the dynamics with endogenous noise – assuming an identical involvement-noise relationship in both populations – are as in the case of asymmetric noise: Both the SPE and a state corresponding to a mixed-strategy NE with symmetric division are asymptotically stable.

Gutsche (2000) criticizes the particular proxy for agents' involvement chosen by Gale et al. (1995). Namely, agents are on the one hand assumed to imitate fellow agents without any evaluation of their comparative success, but on the other hand have an involvement based on the correct expected payoffs of their strategies. Instead, Gutsche investigates a specification in which agents' involvement is proportional to the average payoff of their population. Then, noise levels in states close to component $\mathcal{N}$ are roughly equal in both populations, and – as in the case of symmetric noise – only the SPE is asymptotically stable.

Clearly, many different endogenizations of noise are possible – and produce dynamics either like in Fig. 1.15 or in Fig. 1.16. In any case, the model of Gale, Binmore, and Samuelson (1995) illustrates that the SPE prediction for the 2 × 2-minigame version of the ultimatum game[72] retains a special relevance also in a setting with boundedly rational players. However, it may take long before evolutionary forces induce this highly rational outcome. For some initial population shares and noise dynamics, the SPE may never be reached at all because some partition of the responder population into Y- and N-players which makes H the optimal offer can also be stable. Chapter 2 investigates the UMG under a different behavioural assumption and with a genuine two-agent setting.

1.4.3 Bargaining Automata

The derivation of the SPE of the ultimatum minigame considered in the previous section requires only one step of backward induction. The unique SPE prediction for the Rubinstein bargaining game (cf. Sect. 1.3.2) rests on much more sophisticated arguments and also more demanding assumptions. It can also be supported by a model of boundedly rational agents. Binmore, Piccione, and Samuelson (1998) put forward an evolutionary argument which is based on the two assumptions that, first, players' bounded rationality is captured by the representation of their strategies as finite automata and, second, more complex strategies are generally more costly, e. g. because they are more difficult to learn or to implement.

Machines are a very general abstract computational model. Formally, a machine can be described by a 6-tupel $M = \langle Q, q_0, \mathcal{O}, \chi, \mathcal{I}, \varsigma \rangle$.[73] Q is the set of *states* of machine M, and q_0 denotes its *initial state*. For each state $q \in Q$, an *output* x from the set $\mathcal{O}$ of feasible outputs is specified by the *output function* $\chi \colon Q \to \mathcal{O}$. The set $\mathcal{I}$ describes processable *inputs*. The *transition function* $\varsigma \colon Q \times \mathcal{I} \to Q$ specifies a successor state for any $q \in Q$ given some input in $\mathcal{I}$.

A player's pure strategy $s_i \in S_i$ in an extensive game can elegantly be represented as a machine. One can choose the set of states, Q, to equal the set of i's decision nodes. The output $\chi(q)$ then corresponds to the move specified by s_i for information set q. A transition is induced by input from the opponent player $-i$, i. e. $-i$'s moves. It leads to a new information set, a new move by i, etc., until a terminal state is reached from which no further transitions occur. Such a machine representation implies no loss of generality if the set of states, Q, is allowed to be infinite. Restricting attention to *finite*

---

[72] They also run numerical simulations for finer discrete approximations of the ultimatum game – reporting a particular robustness of divisions that give approximately 20% of the surplus to the responder.

[73] Attention will be restricted to deterministic machines. Computer scientists are particularly interested in the sub-class of *Turing machines*; these are used to define which algorithms or functions are *computable* and which are not.

state machines or *finite automata* – characterized by a finite Q – imposes a bound on the complexity of representable strategies. So, finite automata can serve as a model of bounded rationality.

Binmore et al. (1998) consider the infinite-horizon alternating offers game in which players' are perfectly patient, but there is pressure to reach agreement because negotiations can exogenously break down (cf. p. 42). A finite *bargaining automaton* M produces state-dependent demands $\chi(q) \in X = [0,1]$, which implicitly reject an earlier opponent's offer, as its output, and may accept some demand $x \in X$ which it has received as its input.

It is convenient to let $x(t)$ denote the surplus share which accrues to the *current* proposer of period t if the current responder accepts.[74] Since it is a priori not known whether automaton M will act in the role of player 1 and make the first demand, or whether it will first be called to act on the opponent's demand, the initial state q_0 of any bargaining automaton M has a special feature. Its output $\chi(q_0)$ is only produced if M is exogenously triggered to do so by having Nature assign it the role of player 1. In order to allow an automaton to explicitly condition its behaviour on the role of player 1, q_0 can have the special character I^* as its output instead of a demand $x(0) \in X$. This directly leads to a transition to a new state $\tilde{q} = \varsigma(q_0, I^*)$, where demand $x(0)$ will then be made. Otherwise, input is received and responded to only from the opponent machine. The output Y results if the current responder accepts the current proposal $x(t)$. From any state, the input Y implies a transition to the terminal acceptance state q^*.

Let $\mathcal{M}$ denote the set of all finite bargaining automata, where each $M \in \mathcal{M}$ is defined by its individual set of states $Q_M = \{q_1, \ldots, q_n\} \cup \{q_0, q^*\}$, its output function $\chi_M : Q_M \to X \cup \{I^*\} \cup \{Y\}$ where $\chi_M(q) = I^*$ implies $q = q_0$, and its transition function $\varsigma_M : Q_M \times X \cup \{q_0, I^*\} \cup Q_M \times \{Y\} \to Q_M$ where $\varsigma_M(q, Y) = q^*$. A *machine bargaining game* is played between two bargaining automata $M_I, M_{II} \in \mathcal{M}$. It starts in period $t = 0$ with the random assignment of player roles, where each automaton has equal chance to be player 1. The automaton acting as player 1 will be denoted by $M_{(1)}$ and its opponent by $M_{(2)}$. After Nature's random draw, $M_{(1)}$ makes the first demand $x(0)$, and prompts a chain of offers and counter-offers. As in the original Rubinstein bargaining game, the game moves from period t to period $t + 1$ after each rejection, i.e. any transition that is not induced by I^* and is not to q^*. With probability $p \in (0, 1)$ the game ends in period $t + 1$ before a counteroffer is made, resulting in a zero share of the surplus for both machines. A transition to q^* following a demand $x(t)$ by M_I gives it a surplus share of $x(t)$ and M_{II} receives the share $1 - x(t)$; if $x(t)$ has been made by M_{II}, then M_I receives $1 - x(t)$. Figure 1.17 shows two bargaining automata and the induced sequence

---

[74] The notation $x(t)$ rather than x_t will be used here in order to stress that demands are *not* stated in terms of the share of the initial proposer, player 1, but the current proposer. Nevertheless, periods $t \in T = \{0, 1, 2, \ldots\}$ are discrete.

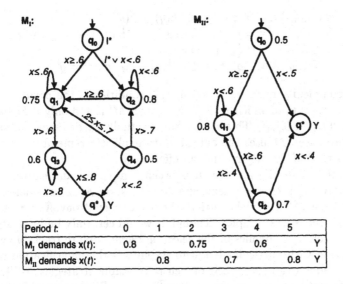

Period t:	0	1	2	3	4	5
M_I demands $x(t)$:	0.8		0.75		0.6	Y
M_{II} demands $x(t)$:		0.8		0.7	0.8	Y

Fig. 1.17. Two bargaining automata and a possible negotiation (Binmore et al. 1998, p. 267; transitions after input Y are omitted)

of offers when M_I is selected as $M_{(1)}$ and no exogenous breakdown occurs before period 5.

The expected surplus share obtained by an automaton $M_I \in \mathcal{M}$ conditional on playing against $M_{II} \in \mathcal{M}$ in the role of player 1 will be denoted by $\pi_{(1)}(M_I, M_{II})$. Similarly, $\pi_{(2)}(M_I, M_{II})$ is M_I's expected surplus share in the role of player 2, and $\pi(M_I, M_{II}) = \frac{1}{2}\pi_{(1)}(M_I, M_{II}) + \frac{1}{2}\pi_{(2)}(M_I, M_{II})$ is M_I's unconditional expected payoff. The expected payoff is one of two determinants of evolutionary success in an encompassing *automaton selection game*. In this game, two symmetric *meta-players*, I and II, are imagined to choose $M_I, M_{II} \in \mathcal{M}$ as their respective strategy for a machine bargaining game based on automata's fitness. This fitness is also affected by the automaton's *complexity*.

Different measures of complexity of a finite automaton exist. Binmore et al. (1998) assume an arbitrary complete complexity relation $\succcurlyeq$ for which $M' \succ M$ holds whenever a *collapsing state condition* is satisfied.[75] This specifies that M is, in particular, less complex than M' if it can be obtained from M' by condensing two states in $Q_{M'}$ into one state in Q_M, or by deleting a state state such as the unreachable q_4 in automaton M_I of Fig. 1.17. More precisely, their condition requires $M' \succ M$ to be true whenever there exists a surjection $h\colon Q_{M'} \to Q_M$ such that

---

[75] $\succ$ is typically incomplete, so that many automata are considered as equally complex.

$$(\exists q \in Q_{M'})\,(\exists q' \neq q \in Q_{M'})\colon h(q) = h(q')$$
$$\wedge\ (\forall q \in Q_M)\colon \left\{x = \chi_M(q) \implies (\exists q' \in h^{-1}(q))\colon\ x = \chi_{M'}(q')\right\}$$
$$\wedge\ (\forall x \in X \cup \{I^*\})\colon \left\{\varsigma_M(q, x) = \tilde{q}\right.$$
$$\left. \implies (\exists q' \in h^{-1}(q))\,(\exists \tilde{q}' \in h^{-1}(\tilde{q}))\colon \varsigma_{M'}(q', x) = \tilde{q}'\right\}.$$

A simple complexity measure satisfying this condition is the *counting-states criterion*, according to which M is simpler than M' if and only if M has less states, i.e. $|Q_M| < |Q_{M'}|$. This means that it actually carries some complexity costs if automaton M uses the output I^* to explicitly remember its role as player 1 instead of directly making an offer in q_0.

It remains to specify a trade-off between complexity and expected payoffs. Binmore et al. choose a lexicographic criterion: M is considered more successful than M' if it yields strictly higher expected payoff or yields the same payoff but is less complex. Though it would certainly be interesting to study evolutionary dynamics of the selection game, a more tractable way to find automata with particular long-run stability is the identification of evolutionary stable strategies (ESS) or neutrally stable strategies (NSS). ESS do typically not exist for extensive games[76] and so Binmore et al. adapt the weaker NSS concept to lexicographic fitness. They call an automaton M_I a *modified evolutionary stable strategy (MESS)* if for all $M_{II} \neq M_I \in \mathcal{M}$

1. $\pi(M_I, M_I) > \pi(M_{II}, M_I)$, or
2. $\pi(M_I, M_I) = \pi(M_{II}, M_I)$ but $\pi(M_I, M_{II}) > \pi(M_{II}, M_{II})$, or
3. $\pi(M_I, M_I) = \pi(M_{II}, M_I)$ and $\pi(M_I, M_{II}) = \pi(M_{II}, M_{II})$ but $M_{II} \succcurlyeq M_I$.

Writing $\delta = 1 - p$ for the probability that the game between two automata continues after a rejected offer, the following is true:

Theorem 1.5. (Binmore-Piccione-Samuelson) *For any MESS $M \in \mathcal{M}$ of the bargaining automata selection game, there is immediate agreement when automaton M plays itself. M's payoff against itself in the role of player 1 is bounded by the proposer's and responder's SPE payoff of the corresponding Rubinstein bargaining game with exogenous breakdown risk, i.e.*

$$\frac{\delta}{1 + \delta} \le \pi_{(1)}(M, M) \le \frac{1}{1 + \delta}. \tag{1.39}$$

Sketch of the proof: The reader is referred to Binmore et al. (1998) for a rigorous proof. Lemma 1 of Binmore et al. (1998) establishes that if M is a MESS and plays itself, then none of its states $\{q_0, \dots, q_n\}$ can be used more then once. For illustration, assume that a MESS M is in the same state q_0 in periods 0 and 2 when playing itself. This means that $M_{(1)}$'s offer $x(0)$ is rejected, $M_{(2)}$'s counter-offer $x(1)$ is also rejected and $M_{(1)}$ stays in q_0. Hence $x(2) = x(0)$, and finally $M_{(2)}$'s offer $x(3) \neq x(1)$ leads to some new

---

[76] The reason is that mutants who deviate in un-reached decision nodes cannot be discriminated. Considering complexity costs may alleviate this problem, but this is not explored by Binmore et al.

state $\tilde{q} \neq q_0$. Clearly, an automaton M' that mimics M's behaviour in the role of $M_{(1)}$, but directly offers $x(3)$ instead of $x(1)$ in the role of $M_{(2)}$ in period 1 receives higher expected payoff – a contradiction to M being a MESS.

The second step is to show that any MESS M which does not condition its behaviour on whether it is player 1 or not, i.e. $\chi_M(q_0) \neq I^*$, must immediately agree on demand $x(0)$ when playing itself. Assume instead that M is a MESS but agreement is reached in $t \geq 1$. Consider an automaton M' that conditions its behaviour on whether it is player 1, and that mimics M's transitions and output as $M_{(2)}$ when it is in fact $M_{(1)}$ – making an initial demand $x(0)$ that is equal to the demand $x(1)$ that M would make after an initial rejection. Also, let M' mimic M's play as $M_{(1)}$ when it is in fact $M_{(2)}$. So M' copies M with switched roles. M' gets the same expected payoff playing M in the role of player 1 which M gets against itself in the role of player 2, but one period earlier. Since M is a MESS, $\pi_{(1)}(M', M) = \pi_{(2)}(M, M)/\delta \leq \pi_{(1)}(M, M)$. Similarly, M' gets the same expected payoff in the role of player 2 that M gets in the role of player 1, but one period later and so $\pi_{(2)}(M', M) = \delta \pi_{(1)}(M, M) \leq \pi_{(2)}(M, M)$. These two inequalities combine to $\delta \pi_{(1)}(M, M) = \pi_{(2)}(M, M)$ and imply that $\pi(M', M) = \pi(M, M)$. However, it can be shown that $\pi(M', M') > \pi(M, M')$. This is a contradiction to M being a MESS, and so a MESS M with $\chi_M(q_0) \neq I^*$ must immediately agree to its own offer.

Binmore et al. then confirm that an automaton with $\chi_M(q_0) = I^*$ is unnecessarily complex. This is done by, first, establishing that different demands $x(0)$ and $x(1)$ must be observed in periods 0 and 1 when any MESS M plays itself. Since different states have to be used to produce $x(0)$ and $x(1)$ as output even if $x(0) = x(1)$ by the first lemma, identical offers in the first two periods either mean that states of M could be collapsed without affecting payoff, or agreement is reached later than possible. Using $x(0) \neq x(1)$ and assuming $\chi_M(q_0) \neq I^*$, the different outputs that are necessarily attached to states q_0 and its immediate successors can serve as an indicator of whether M plays in the role of player 1 or not. Hence, a state q_0 with $\chi_M(q_0) = I^*$ is redundant. So by the second step, agreement must occur immediately if a MESS M plays itself.

Playing against itself, the MESS M demands $x(0)$ in state q_0, and receives the share $x(0)$ if this offer is accepted. Not triggered to produce output $x(0)$ itself as $M_{(2)}$, it moves from q_0 to acceptance state q^* after a demand $x(0)$, receiving the share $1 - x(0)$. If x_0 were too large, it would be profitable for a mutant M' to delay the game by one period whenever it is assigned the disadvantageous role of $M_{(2)}$ and to demand $x(1) = x(0)$ in $t = 1$; M would then (wrongly) assume that itself is $M_{(2)}$, and accept. Such a mutant M' is not profitable if and only if $1 - x(0) \geq \delta x(0)$. Similarly, if x_0 were too small, it would be profitable for a mutant M' to delay the game by one period if it happens to be assigned the role of player 1. This is not profitable if and only if $x_0 \geq \delta(1 - x_0)$. Combining these two inequalities and using $\pi_{(1)}(M, M) = x(0)$ yields (1.39). $\square$

The result of Binmore et al. (1998) holds also when a weakly concave transformation is applied to absolute surplus shares by $\pi_{(i)}$, $i = 1, 2$. This can be interpreted as having symmetrically risk-averse meta-players. The perfect divisibility of surplus $X = [0, 1]$ is not required in the proof of Theorem 1.5. So, in contrast to the unique SPE result of Rubinstein (1982) (cf. the discussion of van Damme et al. 1990 on p. 43), the result also holds when demands have to be made in multiples of a smallest (monetary) unit ξ, i.e. the case of $\chi(q) \in X_\xi = \{0, \xi, 2\xi, \ldots, 1\}$. As the continuation probability δ approaches 1, the SPE payoff bounds imposed by (1.39) become tighter; the fifty-fifty division of the surplus is reached by any MESS playing itself in the limit.

Note that Theorem 1.5 is a characterization result – nothing is said about the conditions under which a MESS exists. Binmore et al. (1998), however, show existence of a MESS for a reasonable restriction of the complexity relation $\succcurlyeq$, which is satisfied e. g. by the counting-states criterion. Then, a MESS which implements the fifty-fifty division exists whenever $0.5 \in X_\xi$.

Theorem 1.5 and the corresponding existence results provide evolutionary support for the unique SPE prediction for the considered variation of a Rubinstein bargaining game. Players need not be capable of complex mathematical calculations; no infinitesimally small monetary unit needs to exist. Nevertheless, existence of a MESS satisfying (1.39) says little about how or whether it would be reached by a dynamic adaptation process. If it is reached, even isolated mutants are not necessarily repelled – not to speak of persistent noise or drift. Also, Binmore et al. demonstrate that Theorem 1.5 cannot straightforwardly be extended to an asymmetric setting. Thus, the SPE's comparative statics for two differently risk-averse or impatient (populations of) players cannot be corroborated in their model.

Taking a slightly different approach, Chatterjee and Sabourian (2000) consider finite automata which play an n-person alternating offers bargaining game. Their game extends the procedure proposed by Shaked for three bargainers (cf. p. 44) to n players and has exogenous breakdown risk instead of impatient players; the game considered by Binmore et al. is the special case of $n = 2$. Chatterjee and Sabourian first investigate *Nash equilibria with complexity costs* (NEC) of the automaton selection game. A NEC is a strategy or automata profile $(M_1, M_2, \ldots, M_n)$ such that each automaton M_i is a best reply to the other automata – denoted by M_{-i} – and no less complex automaton $M_i' \prec M_i$ is also a best reply to M_{-i}. This formalizes a lexicographic trade-off between payoff and complexity as in Binmore et al. (1998). Chatterjee and Sabourian, however, find that a NEC sustains any arbitrary surplus partition with agreement in any of the first n periods. Perpetual disagreement is the only other NEC. All these equilibria involve automata M_i that behave the same way in any 'stage' of n periods, i.e. each automaton makes the same proposal whenever it is its turn to propose. So, NEC non-trivially induces stationarity, but neither immediate agreement nor a restriction of the surplus division.

In contrast, a strong result is obtained when automata are assumed to make errors with arbitrarily small probability: Any NEC in the game played with noisy automata involves immediate agreement on the unique stationary SPE payoffs of the underlying n-person alternating offers bargaining game. For the special case of $n = 2$, this implies immediate agreement on the SPE surplus division $(\frac{1}{1+\delta}, \frac{\delta}{1+\delta})$ – establishing yet more support for Rubinstein's SPE result by a model with error-prone boundedly rational players.

Chatterjee and Sabourian use a complexity criterion different from those considered by Binmore et al. and, in particular, the counting states criterion. However, if the counting states criterion is used in their setting, the results remain true for $n = 2$ (not $n > 2$). Because Chatterjee and Sabourian consider entire NEC strategy profiles $(M_1, M_2, \ldots, M_n)$, and do not restrict attention to single MESS strategies that play themselves, each automaton M_i is conditioned on its role as player i without any complexity costs. An analogous version of the step in the proof of Theorem 1.5 which has ruled out explicit conditioning in a MESS therefore cannot be applied in Chatterjee and Sabourian's setting. This causes the less sharp results when Chatterjee and Sabourian consider unperturbed NEC as compared to Binmore et al.'s MESS characterization.

1.5 Empirical Evidence and Discussion

Several theoretical approaches to bilateral bargaining have been presented in this chapter (cf. Fig. 1.18 for a selective overview), starting with Edgeworth's early formalization of individual and collective rationality in a bargaining situation, Zeuthen's assumptions on concession behaviour under risk, and Hicks' equilibrium of strike resistances, i. e. anticipated costs of disagreement. Nash's solution via explicit axioms for a 'reasonable' agreement given a well-defined bargaining problem nicely matches Zeuthen's predictions, and highlights the role of risk preference and the status quo. It is itself corroborated by the SPE analysis of Rubinstein's extensive bargaining game and its modifications, which allow for explicit analysis of time preference, outside options, breakdown risk, and sequencing of players' decisions. The corresponding predictions can be confirmed without the very demanding assumptions on players' rationality underlying SPE analysis. This is demonstrated by Young's model of adaptive play with myopic responses to partially sampled information, the simple imitation heuristic considered by Gale et al., or the automaton model of Binmore et al. So, for the case of complete information, several complementing models of bargaining establish a rather coherent theory – each model having a different focus, but typically predicting something that tends towards the asymmetric Nash solution.

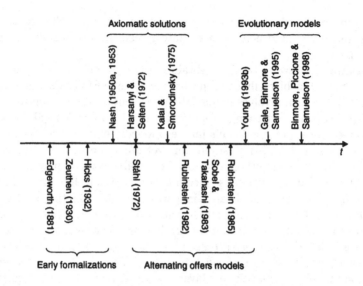

Fig. 1.18. Selected contributions to bargaining theory

This theory can be regarded as a goal in itself. It is also a prescriptive tool, and can be used to calculate strategies or to direct the pre-bargaining design of institutions and procedures. For example, the effects on players' respective bargaining power of the sequencing of moves – the advantage of the initial and final proposer and possibly asymmetric time intervals between two offers – or known fixed costs at different stages of bargaining can be analysed a priori. The roles of player characteristics like risk-aversion, time-preference, outside options, processing capabilities, and also information asymmetries are clarified.

It is of crucial importance, though, whether above models do accurately capture all driving forces of bargaining and if the theory can serve as a descriptive and predictive tool. For example, do people really acknowledge and exploit the extreme theoretic bargaining power of the proposer in ultimatum game situations? The obvious way to answer this question is empirical research. Unfortunately, bargaining situations that are as clearly defined as above and have exogenously given procedures are rare in reality. Hence a rejection or corroboration of bargaining theory based on field data seems extremely hard. Economists have mostly resorted to testing the point predictions and qualitative implications in laboratory experiments. Quite detailed overviews of these studies are given by Camerer and Thaler (1995), Güth (1995), Roth (1995a), and Roth (1995b).

One of the earliest such experimental investigations is described by Nydegger and Owen (1974). They study thirty pairs of students who bargain face to face over the division of a dollar or a certain number of poker chips. Giving no structure at all for one group of ten pairs, imposing an upper bound of 60 cent on the share of player 1 in the second group, and using player-specific chip-to-dollar exchange rates in the third group, Nydegger and Owen test the axioms underlying the Nash solution (Sect. 1.2.1) and the Kalai-Smorodinsky solution (Sect. 1.2.2). A fifty-fifty division of the money turns out to be agreed on by all 30 pairs. This observation questions the Kalai-Smorodinsky monotonicity axiom (MON), and the invariance to equivalent utility representations (INV) which is common to both bargaining solutions.

Alvin Roth and collaborators, too, have carried out similar experiments without a fixed bargaining protocol. Their experiment design allows communication only via anonymous computer terminals and, importantly, controls for possible asymmetries in subjects' von Neumann-Morgenstern utility for money. Namely, participants bargain over the division of lottery tickets. After the negotiations, one player $i = 1, 2$ is drawn to win a player-specific large prize x_i^L. Winning chances are proportional to ticket shares. The loser, $-i$, receives a small prize x_{-i}^S which is equal to his payment in case of no agreement, x_{-i}^D. Since expected utility is linear in probability, this creates a perfectly symmetric setting and allows to make players' preferences in the bargaining situation common knowledge.[77] According to the Nash or Kalai-Smorodinsky solutions, it should not matter what the player-specific prizes are and whether both players know them. Roth and Malouf (1979) find, however, that common knowledge about unequal prizes matters a lot, and makes the ticket division which yields equal expected monetary payoff to both players a focal point for agreements. This observation is confirmed by Roth and Murnighan (1982). A considerable increase in the frequency of disagreement is observed when only the bargainer with lower prize knows both prizes. This casts doubt on the descriptive power of axiomatic models based on von Neumann-Morgenstern utility.

Murnighan, Roth, and Schoumaker (1988) test whether the qualitative predictions of these solutions are nevertheless valid. In particular, they investigate the effect of players' risk-aversion on their share of lottery tickets when the loser's small prize x_{-i}^S is different from the respective disagreement prize x_{-i}^D. The theoretical prediction for symmetric prizes $x_1^S = x_2^S = x^S$, $x_1^L = x_2^L = x^L$, and $x_1^D = x_2^D = x^D$ is that the more risk averse of the players[78] receives less than half the lottery tickets if $x^L > x^D$ and more than half if $x^L < x^D$ (cf. p. 20). A small effect of risk aversion in the predicted direction is observed. But only the weaker theoretical prediction that the more risk averse player, say player 1, fares better in games with $x_1^L < x_1^D$ than in games with $x_1^L > x_1^D$ can be statistically confirmed. Roth (1995b, p. 48) concludes

---

[77] This is based on the two implicit assumptions that subjects' preferences satisfy the von Neumann-Morgenstern axioms, and that this fact is common knowledge. Many decision experiments – collected and complemented with an overview of alternative assumptions e. g. by Lopes (1996) – question this.

[78] Risk aversion is measured in an initial experimental stage in which players have to take several decisions under risk.

> ... it has so far proved far easier to observe the unpredicted effects of
> information [about prizes that do not enter the Nash solution] than the
> predicted effects of risk aversion on the outcome of bargaining ... [Based]
> on the evidence so far available, we cannot deliver a conclusive verdict on
> the overall health of every aspect of theories of bargaining such as Nash's.

Güth, Schmittberger, and Schwarze (1982) are the first to test bargaining behaviour in an explicit extensive game, the ultimatum game (Sect. 1.3.1). They find that proposers do not demand approximately the entire surplus, but on average less than 70%. While 'fair' fifty-fifty divisions are accepted, a significant share of proposals – almost one fifth in total – is rejected by responding players. Experienced proposers generally ask for more than inexperienced subjects, but receive a lower payment on average.

Initially, there has been controversy about the robustness of Güth et al.'s results. Binmore, Shaked, and Sutton (1985) conduct experiments on the 2-stage alternating offers game (Fig. 1.8 (b)), and make observations that seem more in line with the theory. In particular, they ask subjects who have filled the role of player 2 in a first game to play another game in the role of player 1. They find that the fifty-fifty division is replaced as the first game's modal proposal by demands close to the SPE in the second game (based on the assumption that players maximize monetary income). So, they conjecture that subjects propose an equal division as the default for unknown bargaining problems, but quickly understand the strategic structure of the game and then exhibit "a strong tendency to play 'like a game-theorist'" (Binmore et al., p. 1179). Neelin, Sonnenschein, and Spiegel (1988) duplicate the result of Binmore et al. for the 2-stage game. However, they reject a similar hypothesis concerning 3- and 5-stage games.

These seemingly incoherent observations are partly reconciled by Ochs and Roth (1989). They investigate 2- and 3-period alternating offers games, each with four different discounting arrangements for monetary payments. Based on a considerable number of observations, they identify fairly robust general patterns: The SPE – calculated under the assumptions of individual income maximization and the proposer knowing this – fails as a point predictor for proposals even in the tenth iteration of the game (each time played with a different opponent to avoid repeated-game effects). Some adjustment towards the SPE occurs, but there is a persistent bias in the direction of equal division. A significant proportion (16%) of initial offers is rejected. Strikingly, more than 80% of the rejecting responders make counterproposals that actually ask for less cash than has been rejected.[79] Subjects seem to have an aversion to being treated inequitably. Comparing mean initial demands for different discounting arrangements, the null hypothesis that the SPE does no better than coin flipping as a predictor of the direction of differences is rejected, though not convincingly. Ochs and Roth conclude that theory is a poor predictor both quantitatively and qualitatively.

---

[79] Due to the discounted payments, these offers correspond to a more equitable distribution of a smaller surplus.

A cross-cultural study by Roth, Prasnikar, Okuno-Fujiwara, and Zamir (1991) compares experiments on the ultimatum game and a multi-person market game in Jerusalem, Ljubljana, Pittsburgh, and Tokyo. Behaviour in the market game quickly adjusts to the theoretical prediction of the single player on the short side receiving the entire surplus. In the ultimatum game, the modal demand ranges from only 60% (Jerusalem) to 50% (Pittsburgh and Ljubljana) even in the tenth round; between 13% and 23% of demands are rejected. Studies by Slonim and Roth (1998) and Cameron (1999) suggest that this is robust even when the stakes are high. They vary the monetary surplus by a factor of 25 in the Slovak Republic and even 40 in Indonesia (reaching three times the monthly expenditure of the average participant). Rejections of proportionally equivalent demands become less frequent for bigger stakes, but this effect is not significant in the first iteration of the game. Proposer behaviour is largely invariant to the stakes in both studies. There seems to exist a universal tendency to make rather equitable demands and to sacrifice substantial amounts of surplus in order to punish greedy demands, which has been left out of the theoretical models.

On first view, observations suggest an intrinsic preference for playing fair. But this hypothesis is not supported by equilibrium play in the market game of Roth et al. (1991). Prasnikar and Roth (1992) dismiss the fairness hypothesis for proposers with similar findings for a sequential public good provision game. Harrison and McCabe (1996) observe quick convergence to the proposal and acceptance of highly inequitable offers when the aggregate history of strategy choices in a symmetrized ultimatum game is made available to the subjects – in particular so, if this history is manipulated by 'unfair' computer players. Güth and van Damme (1998) conduct ultimatum bargaining experiments in which a third, entirely inactive player turns out to be assigned only a marginal surplus share by the proposer, and conclude (p. 242): "The experimental data clearly refute the idea that proposers are intrinsically motivated by considerations of fairness." Responders, on the other hand, incur considerable costs in order to avoid being treated unfairly in most studies. Since many proposers anticipate this, fairness is relevant. Güth, Huck, and Ockenfels (1996) even find informed proposers in a 3-player, asymmetric-information variation of the ultimatum game to intentionally "pretend fairness by 'hiding behind some small cake'" (p. 600).

Various attempts have been made to incorporate different types of fairness preference into players' utility functions – efforts associated with the term "neoclassical repair shop" by Güth (1995, p. 342). Although strategic behaviour is influenced by the idea of fairness, the strategic environment itself influences what is perceived to be fair. This is highlighted in studies by Binmore, Morgan, Shaked, and Sutton (1991) and Binmore, Swierzbinski, Hsu, and Proulx (1993). In the former, strategically different alternating offers games with similar a priori predictions concerning the 'fair' outcomes are considered. Observations are clearly biased towards the theoretical prediction and, significantly, subjects' responses to a questionnaire after the experiments, asking what they felt to be fair splits in the different settings, are analogously biased. In the latter study, subjects are in a first experimental phase conditioned on different focal points – corresponding to different axiomatic solutions – in playing the Nash demand game against a computer. In the second phase, they anonymously play other subjects. Not all focal points turn out to be equally stable, but most subjects later report the particular equilibrium reached in their group to be close to "the fair amount" (pp. 395ff). Roth (1995a, p. 271) points out a "chicken and egg problem" which makes pure fairness explanations of observations problematic.

Güth (1995, p. 339) observes:

> After all the ultimatum game appears at first sight like a simple distribution task where both parties should receive an equal share.

Similarly, Binmore, Shaked, and Sutton (1989, pp. 757ff, italics in the original) remark for a more complicated bargaining game with outside options:

> We do not ... believe that our subjects ... are gifted with the capacity for effortless mental arithmetic. Without extensive opportunities for trial-and-error learning, they can only be anticipated to have a *dim* awareness of the strategic realities.

It seems plausible that people need time to learn how to play a game. So a number of learning models have recently been investigated not only in the context of bargaining (cf. Sects. 1.4.1 and 1.4.2), but also for other classes of games. Roth and Erev (1995) study a model of entirely introspective reinforcement learning in the ultimatum game, the market game of Roth et al. (1991), and the game of Prasnikar and Roth (1992). The comparison of computer simulations and actual observed dynamics in the ten experiment iterations suggests that their simple adaptive model captures an important aspect of bargaining behaviour. In other games, people are observed to primarily respond to counterfactual, forgone payoffs and to rationally update expectations. So the behavioural model of Camerer and Ho (1999), which combines pure reinforcement learning with belief learning, has received considerable attention. Camerer and Ho do not fit their model to bargaining data and so it is presently not known how relevant their setting is in this context. For alternating offers games with several stages Roth and Erev's model also remains to be evaluated.

In any case, it would be wrong to infer from above studies that (Cameron 1999, p. 47)

> ... the standard game theory predictions are strongly falsified by experimental evidence.

The reason is that game-theoretic predictions for the ultimatum game and other sequential bargaining games have not properly been tested so far. Only the joint hypothesis that

1. each players' objective is to maximize his monetary income and this fact is mutual or even common knowledge to the subjects and
2. players are sequentially rational in the sense of subgame perfectness and this fact is mutual or common knowledge

has been tested and rejected by the experiments. The uncontrolled elements in bargainers' utility and epistemic state are only noted in passing – if at all – by many experimenters. This has recently sparked the critique of Weibull (2000). He suggests how proper tests can perhaps be designed.[80] Earlier, Kennan and Wilson (1993, p. 93) have remarked that "private knowledge enters through the back door" in the above experiments, implying that the SPE is not the correct theoretic benchmark.

The mentioned experimental results should cautiously be interpreted in two ways. First, the persistent divergence between observations and the supposed theoretical predictions indicates that what may at first sight be modelled as a bargaining situation is in fact none in the eyes of the players. Second, observed dynamics of behaviour as subjects gain experience with a game indicate that adaptation and learning take place. This process has to be given still more attention.

The serious methodological issues involved in testing bargaining theory are only an indication of the complexity of real-life bargaining. In view of many plausible and useful implications of bargaining models, but also in view of experimental research that questions too straightforward conclusions, it is apt to close the chapter with the dogma of modelling as expressed by Dodds (1973, p. 310, italics in the original):

(1) Never *believe* a model, and
(2) never *ignore* a model.

---

[80] See Binmore, McCarthy, Ponti, Samuelson, and Shaked (1999) for an effort to separate rationality and preference issues in ultimatum game and 2-stage alternating offers game experiments.

2. Aspiration-based Bargaining

This chapter combines stochastic evolutionary methodology (cf. Sect. 1.4.1) with satisficing behaviour in a bargaining situation, and analyses the latter's implications for efficiency and distribution. Two boundedly rational agents recurrently play the ultimatum minigame (see Sect. 1.4.2) in fixed roles. Each player *satisfices*, i.e. sticks to his past action if it has been satisfactory relative to an endogenous aspiration level; otherwise he abandons it with positive probability. *Aspirations* reflect evolving wishes or heuristic goals which players use for an ex post evaluation of their choice. Players' aspirations rise after positive feedback, i.e. a payoff above the present aspiration level, and fall after negative feedback. Occasionally, aspirations are perturbed.

This satisficing behaviour defines an ergodic Markov process. Its stationary distribution is investigated in order to characterize the long-run implications of the specified dynamics. Satisficing based on endogenous aspiration levels is shown to be 'rational' enough to yield an approximately efficient average outcome when perturbations occur rarely and aspirations are adapted slowly. In the limit, all probability is concentrated on the efficient conventions, i.e. states in which actions result in an efficient surplus division and where additionally players' aspirations equal their respective share. However, the satisficing heuristic does typically not serve to select between the subgame perfect equilibrium and the symmetric Nash equilibrium of the game.

The average surplus distribution is shown to depend on the supports of the perturbation distributions even for vanishing perturbation probability and almost static aspirations. The parameter dependence of the time shares of different surplus-sharing conventions is used to establish novel links between a player's 'character' and his average success in bargaining. The observations provide a behavioural complement to the comparative statics conclusions for completely rational interaction in Sects. 1.2 and 1.3. In particular, Monte-Carlo simulations demonstrate that a player who adapts aspirations more slowly, ceteris paribus fares better. Similarly, greater inertia in changing actions in response to the major dissatisfaction resulting from bargaining disagreement is beneficial. This has the interpretation that it pays to be of persistent and stubborn character. However, there is also a benefit from being capricious in the sense of experiencing big and frequent perturbations in aspirations.

Section 2.1 briefly discusses the related literature. Section 2.2 then presents the satisficing model. Section 2.3 contains the theoretical results, and Sect. 2.4 numerically compares different parameter scenarios. Section 2.5 discusses promising extensions to the model, before Sect. 2.6 collects the proofs of the main results. The presentation draws heavily on Napel (2000).

2.1 Related Literature

Below model is to the author's knowledge the first to combine stochastic evolutionary methodology with satisficing behaviour in a bargaining situation. It draws on the work of Karandikar, Mookherjee, Ray, and Vega-Redondo (1998), who investigate two boundedly rational agents who repeatedly play a fixed symmetric 2×2-game. They find satisficing based on endogenous aspiration levels sufficient for approximate long-run efficiency. In particular, the symmetric Pareto-efficient outcome is selected even if the corresponding strategies are strictly dominated. Since ultimatum bargaining is a fundamentally asymmetric game, Karandikar et al.'s result cannot be directly transferred. It turns out, however, that approximate efficiency is in fact reached in case of the asymmetric ultimatum minigame, too. Yet, more detailed information about the players' adaptation heuristic is required to derive average surplus shares.

The satisficing assumption with endogenous aspiration levels has a long tradition, and has been prominently formulated by Simon (1955, 1959). Simon's work has been elaborated on e. g. by Sauermann and Selten (1962) and Selten (1998), while Winter (1971) provides an early stochastic formalization of a satisficing process. Lant (1992) carries out an explicit empirical investigation of aspiration level adaptation of managers and MBA students in a marketing strategy game. Her study highlights that below simple exponential smoothing rule for aspiration updating is a simplification of real satisficing behaviour. Subjects in Lant's experiments, for example, exhibit an optimistic bias in their aspiration updating and sometimes "over-shoot", meaning that new aspirations may lie above (below) the maximum (minimum) of old aspiration and payoff. Since Lant has asked her participants about their "sales objective" for the next period, the identification of actual aspirations is complicated by the underlying forecasting problem as well as a conceptual difference between subjects' (optimistic) goals and later (realistic?) assessment of success or failure which drives satisficing behaviour. As investigated by Gilboa and Schmeidler (1996), aspiration dynamics can, in fact, be a viable optimization heuristic with very low information processing and calculation requirements, though this is put into a more sceptical perspective by Börgers and Sarin (2000).

Different versions of satisficing based on endogenous aspiration levels in an interactive setting have been investigated by Pazgal (1997), Kim (1999), and Dixon (2000). They are concerned with common interest games in which –

in contrast to bargaining situations – no conflict as to what Pareto-efficient outcome should be chosen can arise. Their findings are similar to those of Karandikar et al. (1998)

The investigation of satisficing players is still comparatively rare in the stochastic evolutionary literature. Most authors consider dynamics either based on myopic best replies (Kandori, Mailath, and Rob 1993, and Young 1993a, 1993b), imitation (e. g. Robson and Vega-Redondo 1996, Hehenkamp 1999, or Josephson and Matros 2000), or possibly a combination (e. g. Kaarbøe and Tieman 1999). Myopic best-response behaviour is already 'too rational' to support the observation of a symmetric surplus division in the ultimatum minigame in the long run: In Young's adaptive play (cf. Sect. 1.4.1), the m-fold repetition of the strict NE corresponding to an asymmetric surplus division is the only recurrence class of unperturbed adaptive play, and hence the only SSS.[1]

A main result of Gale et al. (1995) (cf. Sect. 1.4.2) is that the specification of noise in the proposer and responder population is critical for long-run dynamics. In the absence of noise or when noise in the populations is symmetric, the unique asymptotically stable equilibrium corresponds to an asymmetric surplus division. If the noise level in the responder population is sufficiently greater than in the proposer population, then a mixed responder population combined with equitable offers by (in the limit) all proposers also becomes asymptotically stable. In below model, the specification of noise is important, too. Provided that supports are wide enough, Gale et al.'s finding that more noisy behaviour can benefit a population is confirmed. As the biggest difference to Gale et al.'s quasi-deterministic model, a symmetric division can (in the limit) be the unique long-run outcome independent of the initial state in the stochastic aspiration adaptation model.

Other adaptive learning or evolutionary models of ultimatum bargaining which support the symmetric surplus division include Roth and Erev (1995), Harms (1997), Huck and Oechssler (1999), Peters (2000), and Poulsen (2000). Roth and Erev simulate simple adaptation rules, and obtain predictions close to observations in actual ultimatum game experiments. Harms simulates different symmetrized and discretized versions of the ultimatum game in a replicator dynamics setting, and observes convergence to the symmetric division for a non-negligible set of initial states. Huck and Oechssler analytically investigate a symmetrized ultimatum game, and find that an endogenous preference to punish greedy proposers can evolve – inducing symmetric divisions. Peters uses maximal invasion barriers for NSS in the ultimatum game as his equilibrium selection criterion. He finds that the fifty-fifty division has the highest such barrier given monotonic responder strategies. Poulsen intro-

---

[1] The symmetric division can probably be supported when imitation dynamics are considered. For these – but also for satisficing dynamics based on random exogenous aspiration levels as in Sect. 1.4.2 – the close mathematical relationship to replicator dynamics (cf. e. g. Benaim and Weibull 2000) suggests results similar to the findings of Gale et al. (1995).

duces a share of optimizing proposers, who obtain correct information about responders with positive probability, into the replicator model of Gale et al. (1995); the asymmetric SPE loses its asymptotic stability, while states with symmetric proposals gain stability. A symmetric division also features prominently in Ellingsen's (1997) analysis of Nash demand bargaining. In contrast, the computer simulations of finite-horizon alternating offers bargaining by Van Bragt, Gerding, and La Poutré (2000) clearly favour the asymmetric SPE division.

Bergin and Lipman (1996) suggest that stochastic evolutionary equilibrium selection results based on state-independent perturbation rates have to be interpreted with caution: *any* stationary distribution of the unperturbed process can be 'selected' by the perturbed process if perturbation rates are *state-dependently* chosen in an appropriate way. The author's intuition is, though, that a highly implausible type of state-dependent perturbation probability would have to be chosen in below model to ensure positive weight on the inefficient disagreement outcome in the limit. This intuition is corroborated by findings of Kaarbøe and Tieman (1999). But as will be seen in Theorem 2.2, state-dependent perturbation *supports* drive a non-selection result concerning different efficient surplus distributions in a repeated bargaining situation.

2.2 The Model

The *ultimatum minigame* (UMG) has been discussed in Sect. 1.4.2 in the context of Gale, Binmore, and Samuelson's (1995) imitation model. Its extensive form is shown again in Fig. 2.1. Two players can share an available surplus of four units provided that they are able to agree on how exactly it is to be split. Both players are assumed to linearly value only their own share – the bigger it is, the better to them. The surplus need not be monetary. Habitual bargaining involving e. g. boss and secretary, wife and husband, colleagues, etc. about what work to do, what film to watch, at what time to meet, etc. is a possible example for the model.[2] People frequently seem to not recognize playing games in practice, justifying below behavioural assumptions. The UMG may stand for an entire class of games that from the player's point of view are equivalent in terms of payoffs and aspirations. The strict equilibrium (L, Y) is the unique SPE of the game, but the symmetric surplus distribution implied by Nash equilibrium (H, N) or maximin strategy profile (H, Y) is a focal point in laboratory experiments.[3]

---

[2] The high-stake ultimatum game experiments reported in Sect. 1.5 indicate that boundedly rational behaviour need not be restricted to issues of minor importance.

[3] Yang, Weimann, and Mitropoulos (1999), among others, confirm that subject's behaviour in the ultimatum minigame is just as described in Sect. 1.5 for the ultimatum game.

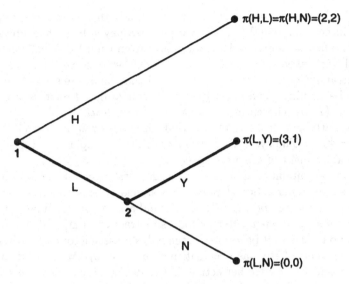

Fig. 2.1. The ultimatum minigame (UMG)

Two agents who recurrently play the UMG in fixed roles are considered. Each interaction is indexed by some $t \in \{0, 1, 2, \ldots\}$ which is referred to as the *period* or *time*. Players are assumed to satisfice; they do not apply any dynamic optimization. There is no anticipation and discounting of future payoffs. Hence, the real time elapsing between two interactions need not be equidistant. Both agents use a simple *satisficing heuristic*, which says:

1. Stick to your action if it performed well (relative to an endogenous aspiration level).
2. Otherwise change your action with positive probability, but not certainty.

This heuristic is myopic and, in particular, involves neither backward induction by player 1 nor even simple payoff maximization by player 2. Therefore, nothing would be lost by considering the normal form version of the UMG as Gale et al. do.

Players' aspirations rise after positive feedback, i.e. a payoff above the aspiration level, and fall after negative feedback. Occasionally, aspirations are perturbed. A random shock to a player's aspiration level can be interpreted as the result of positive or negative experience in a different (unmodeled) game, the observation of outcomes in a UMG played by different players, or merely an idiosyncratic optimistic or pessimistic shift in the player's perception of the world. This satisficing heuristic has first been investigated by Karandikar et al. (1998) in the context of symmetric 2 × 2-games. With more specific assumptions, it defines a discrete-time *Markov process*.

It is assumed that players know their respective maximum and minimum feasible payoff in the game, and this is imposed as a bound on their aspiration levels. In any period, both players recall only their own strategy in last period's interaction together with their personal payoff from that interaction. There is no further individual memory apart from what has been 'condensed' into a player's aspiration level. Agents do neither observe their opponent's payoff, strategy, or aspiration level nor need they be aware even of playing a 2-person bargaining game at all. Player 1's *state* at date t refers to her action $s_1^t \in S_1 = \{L, H\}$ chosen in t and her *aspiration level* $l_1^t \in L_1 = [0,3]$ held in t.[4] Similarly, player 2's state in period t is given by $s_2^t \in S_2 = \{Y, N\}$ and $l_2^t \in L_2 = [0,2]$. The system's state in t is thus a 4-tuple $x_t = (s_1^t, l_1^t, s_2^t, l_2^t)$ which is an element of the state space $E = S_1 \times L_1 \times S_2 \times L_2$.

Actions are updated as follows. Consider first player 1. The probability that she sticks to her action s_1^t in $t+1$ is assumed to be a (weakly) decreasing function p_1 of the gap between her actual and aspired payoff in period t. Formally, her *disappointment* after period t's interaction is $\Delta_1^t = l_1^t - \pi_1(s_1^t, s_2^t)$, which is non-positive if payoff has actually been satisfactory. She will repeat s_1^t with certainty if she has been satisfied in t, i.e. $p_1(\Delta_1^t) = 1$ for $\Delta_1^t \leq 0$. Otherwise she may switch her action. She may also play s_1^t again, even if she has been maximally dissatisfied. This means that function p_1 decreases but is bounded below by some $\tilde{p}_1 \in (0,1)$ – reflecting that player 1's behaviour is subject to *inertia*.

Function p_1 is assumed to be continuous. Moreover, the rate at which p_1 falls is bounded, i.e. there is some $K_1 < \infty$ for which $p_1(0) - p_1(\Delta) \leq K_1 \Delta$ for any Δ. Finally, it is assumed that p_1 falls at least linearly for small dissatisfaction, i.e. there exist $k_1 > 0$ and $\check{\Delta} > 0$ so that $p_1(0) - p_1(\Delta) \geq k_1 \Delta$ for $\Delta \in (0, \check{\Delta})$. The former assumptions formalize that willingness to switch behaviour is changing gradually and at moderate rate. The latter assumption formalizes that players' propensity to stick with the past action decreases visibly as soon as a player is dissatisfied.

Player 2 behaves analogously, but according to a possibly different function $p_2 \colon [-2, 2] \to (\tilde{p}_2, 1]$ for some $\tilde{p}_2 \in (0, 1)$. Functions p_i will be referred to as *inertia functions*; subscripts on Δ are generally omitted. A possible inertia function p_i is illustrated in Fig. 2.2. Action updating is assumed to be independent across time and players for any given state.

Concerning the updating of aspirations, two cases are distinguished (cf. the adaptive play process, Sect. 1.4.1). The first is unperturbed aspiration adaptation; the resulting Markov process will be referred to as the *unperturbed satisficing process* Φ^0. In Φ^0, the aspiration level for $t+1$ is simply a weighted average of the aspiration level in t and the payoff experience in t. More precisely,

$$l_1^{t+1} = l_1^*(x_t) := \lambda l_1^t + (1 - \lambda)\pi_1(s_1^t, s_2^t) \qquad (2.1)$$

---

[4] Some might prefer to call s_1^t player 1's *behavioural mode* in period t since she does not actively 'choose' any action or strategy in the traditional way.

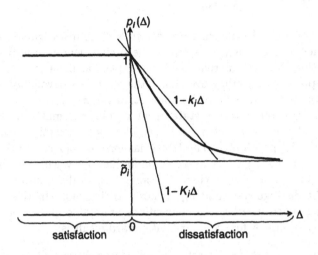

Fig. 2.2. A possible inertia function p_i, specifying the probability of player i sticking with the present action in response to dissatisfaction Δ

with $\lambda \in (0, 1)$ for player 1. Player 2's aspirations are exponentially smoothed analogously and, for the time being, with the same aspiration persistence parameter λ defining the aspiration update function l_2^*.

In case of perturbed aspiration adaptation, both player's aspirations are first deterministically updated as in Φ^0. Then, each player may independently experience a *perturbation* in aspirations with a small probability. The new aspiration level l_i^{t+1} is a random variable whose distribution can depend on the deterministically updated 'intermediate' aspiration level $l_i^*(x_t)$.

To facilitate theoretical analysis, both players are assumed to have the same state-independent probability $\eta > 0$ of experiencing a tremble. Actual perturbations are distributed differently for player 1 and 2. Individual post-perturbation aspirations l_i^{t+1} are assumed to have the continuous conditional density $g_i(\cdot | l_i^*(x_t))$ with its support contained in L_i. Moreover, it is assumed that whenever player i experiences a perturbation, his resulting aspiration level l_i^{t+1} has positive probability of staying close to deterministically updated aspirations $l_i^*(x_t)$; also, intermediate aspirations may jump back from $l_i^*(x_t)$ to l_i^t and at least some $\varepsilon > 0$ further.[5] The resulting *perturbed satisficing process* will be denoted by Φ^η.

---

[5] A sufficient requirement for player 1 is that $g_1(l_1^{t+1} | l_1^*(x_t)) > 0$ for all $l_1^{t+1} \in [l_1^*(x_t) - 3(1 - \lambda) - \varepsilon,\ l_1^*(x_t) + 3(1 - \lambda) + \varepsilon] \cap L_1$. The assumption is not necessary, but guarantees that Φ^η is open set irreducible, i.e. every open subset $Y \subset E$ is accessible from any $x \in E$. An alternative modelling assumption is that aspirations are *not* deterministically updated at all in case of a perturbation.

2.3 Theoretical Results

First, the unperturbed satisficing process Φ^0 will be investigated, and then Φ^η is considered. Capturing the spirit of the similar definition in Sect. 1.4.1, but accounting for the different type of adaptation behaviour, a state in which both players' aspiration levels equal the respective payoffs will be called a *convention*. Let $c_{s_1 s_2}$ denote the convention $(s_1, \pi_1(s_1, s_2), s_2, \pi_2(s_1, s_2))$. For example c_{HN} refers to the state in which player 1 makes a high offer, player 2 would reject a low offer, and both players are exactly satisfied with their (symmetric) payoff of 2. The UMG has exactly four conventions – one corresponding to each pure strategy combination. They will be collected in the set C. As c_{HY} and c_{HN} yield the same surplus distribution, they are typically not distinguished and $c_{H.}$ denotes their union. In the following, it will be examined whether the conventions are stable under the specified satisficing dynamics. A first and straightforward result is:

Proposition 2.1. *Let the unperturbed satisficing process Φ^0 be defined as above. State $x \in E$ is an absorbing state of Φ^0 if and only if it is a convention, i.e. $x \in C$.*

Proof. First, Φ^0 never leaves a convention $c \in C$, which follows directly from the assumption $p_i(0) = 1$ $(i = 1, 2)$ and the aspiration update rule. Second, assume that Φ^0 is in state $x \notin C$ in period t. At least one player – say player 1 – receives less or more than she aspires to, and (2.1) then implies $l_1^{t+1} \neq l_1^t$. So x cannot be absorbing. $\qquad\qquad\square$

It can be shown that the unperturbed satisficing process Φ^0 converges almost surely to a convention:

Proposition 2.2. *Let Φ^0 be defined as above. From any initial state $x_0 \in E$, Φ^0 converges with probability one to a convention $c \in C$.*

The proof is given in Sect. 2.6. Intuitively, from any state x there is a positive probability $\varepsilon > 0$ of starting an infinite run on the present action pair in period t. So the probability of not having started one in the last T periods goes to zero for $T \to \infty$.

Proposition 2.2 establishes that players' adaptive interaction will in the long run settle to a convention, which can also adequately be referred to as an *adaptive equilibrium*. However, any of the three possible bargaining results – symmetric division, asymmetric division, and break-down – can be selected as the long-run outcome of Φ^0 depending on the initial state.

When players experience trembles in their aspiration levels, the influence of the initial state x_0 is gradually washed away. Φ^η is an *ergodic* Markov process:

Proposition 2.3. *Let the perturbed satisficing process Φ^η be defined as above. For any given perturbation parameter $\eta \in (0, 1]$, Φ^η converges (strongly)* *to a unique limit distribution μ^η which is independent of the initial state $x_0 \in E$.*

The proof is given in Sect. 2.6. Proposition 2.3 establishes that Φ^η's long-run behaviour is accurately described by its stationary distribution μ^η. Both empirical frequency distributions over states up to a period t, as sampled from an arbitrary process realization, and the theoretical t-step distribution over state space E given an arbitrary initial state $x_0 \in E$ converge to μ^η as $t \to \infty$.

It is generally not possible to give details on μ^η for arbitrary parameters η and λ. The theoretical investigation will focus on the analytically tractable benchmark case in which the probability of a tremble, η, is close to zero and in which, additionally, present payoff experience affects players' aspirations only marginally, i.e. when λ is close to one.[6] The simulations in Sect. 2.4 confirm that typical dynamics are not very far from this benchmark. The first main result is:

Theorem 2.1. *Let the perturbed satisficing process Φ^η be defined as above. The limit stationary distribution of Φ^η for $\eta \to 0$, μ^*, places positive probability only on the Pareto-efficient conventions c_{LY} and $c_{H.}$ as $\lambda \to 1$.*

The proof is given in Sect. 2.6. Theorem 2.1 establishes that the players will (approximately) divide the available surplus *efficiently* if aspiration trembles are rare and aspirations are adapted slowly.

The intuition for this result is the following: as the probability of a tremble, η, approaches zero, Φ^η becomes more and more like the unperturbed process Φ^0 with extremely rare interrupting shocks. By Proposition 2.2, one therefore knows that Φ^η will spend most of its time in a convention.[7] However, the inefficient convention c_{LN} is unstable: a single perturbation of one player's aspiration level leads directly to convention $c_{H.}$ or c_{LY} with positive probability. This happens because a likely action switch by the now dissatisfied player results in an efficient bargaining outcome. The outcome satisfies both players and therefore the strategy combination is repeated infinitely often – or until another perturbation results in some player's dissatisfaction. Then, however, a move back to c_{LN} is extremely unlikely even if the perturbation causes the play of (L, N): because λ is close to 1, an enormous number of periods would have to pass before the players could again become satisfied with the disagreement outcome. The odds are that at least one of the necessarily dissatisfied players switches his action again and sets the course for approaching an efficient convention.

The second main result is the following:

---

[6] This analysis of limit asymptotic behaviour is greatly facilitated by having a continuous state space E.

[7] Strictly speaking a convention can only in the limit and with zero probability be established from the interior of state space E. An arbitrary *neighbourhood* of the respective state $c_{s_1 s_2}$ is implicitly referred to.

Theorem 2.2. *Let the perturbed satisficing process* Φ^η *be defined as above, and let* μ^* *denote the limit stationary distribution of* Φ^η *for* $\eta \to 0$. *The supports of perturbation densities in conventions* c_H. *and* c_{LY} *can be chosen*

  i) such that μ^* *places all weight on the asymmetric efficient convention* c_{LY},
 ii) such that μ^* *places all weight on the symmetric efficient convention* c_H.,
     or
iii) such that μ^* *places positive weight on both* c_{LY} *and* c_H.,

as $\lambda \to 1$. *In particular, when post-perturbation aspiration densities have full support in* $L_1 \times L_2$, *then* μ^* *places positive weight on both* c_{LY} *and* c_H..

The proof is given in Sect. 2.6. Theorem 2.2 implies that the average *surplus distribution* selected by the satisficing heuristic depends on the distribution of aspiration perturbations even for $\eta \to 0$ and $\lambda \to 1$.

The intuition for this result again rests on the approximative description of Φ^η as being a composition of the unperturbed satisficing process Φ^0 running almost all the time and rare interruptions by a perturbation of one player's aspiration level. Theorem 2.1 established that Φ^η in the limit spends almost all time in c_{LY} and c_H., so only perturbations occurring in these conventions matter. Now suppose there is some small neighbourhood $U(c)$ of either convention $c \in \{c_{LY}, c_H.\}$ such that a perturbation into $U(c)$ leads to c with arbitrarily higher probability than to convention $\bar{c} \neq c$. These neighbourhoods are shown to exist in Sect. 2.6. Then, it is clear that if perturbations from c_{LY} are never leading to aspirations outside of $U(c_{LY})$ and if, in contrast, there is positive probability of a perturbation from c_H. into $U(c_{LY})$, then Φ^η will spend almost all time in the asymmetric convention c_{LY}. Similarly, the stationary distribution is as in ii) when perturbations from c_H. are very 'narrow' and those from c_{LY} are 'wide'. When perturbations have reasonably wide support in both c_{LY} and c_H., iii) applies. Then, the asymmetric and symmetric conventions will alternatingly be in place for long stretches of time, and bargaining dynamics are characterized by *punctuated equilibria*.

Note that the stationary distribution of an ergodic stochastic process only captures average behaviour as time goes to infinity. Considering the limit of such a limit distribution as $\eta \to 0$ and as $\lambda \to 1$, i.e. trembles vanish and persistence of once formed aspirations becomes large, is a useful benchmark. Still, in actual process realizations with parameters close to the limit, the Pareto-inefficient outcome excluded by the limit Theorem 2.1 will be observed. With this caveat, one may summarize above mathematical results as a quite robust prediction concerning efficiency and the demonstration that the specific distribution of surplus – even in the limit – depends on the details of aspiration perturbations.

2.4 Simulation Results

It is worthwhile to ask how stationary distributions of aspirations and actions in Φ^η look like for 'plausible' parameters. Is there a bias in favour of the subgame perfect equilibrium (L, Y)? Moreover, how parameter-sensitive is the average surplus distribution and do monotonic trends exist? A number of parameter scenarios shall be investigated using Monte-Carlo simulation.[8]

The piece-wise linear inertia functions

$$p_i(\Delta) = \begin{cases} 1; & \Delta \le 0 \\ 1 - M_i \Delta; & \Delta \in (0, \frac{1-\tilde{p}_i}{M_i}) \\ \tilde{p}_i; & \Delta \ge \frac{1-\tilde{p}_i}{M_i} \end{cases}$$

with parameters $\tilde{p}_i \in (0,1)$ and $M_i \in \mathbb{R}_{++}$ $(i = 1, 2)$ will be considered. Truncated normal perturbation distributions with mean $l_1^*(x_t)$ and $l_2^*(x_t)$, and player-specific standard deviations σ_1 and σ_2 are used.[9] For more flexibility in modelling different agent 'characters' the assumption that both players have the same aspiration persistence and the same perturbation probability is dropped.

The reference scenario S0 is based on the following parameter choices: $\eta = (0.05, 0.05)$, $\lambda = (0.8, 0.8)$, $\tilde{p} = (0.7, 0.7)$, $M = (1.0, 1.0)$, and $\sigma = (0.1, 0.1)$. In S0, both players on average experience one perturbation in 20 updates, lower their aspiration by 20% after a round of bargaining disagreement, and stick to their action with at least a probability of 0.7.[10]

Typical dynamics of Φ^η are characterized by long stretches of time spent in one efficient adaptive equilibrium, which is occasionally challenged and eventually replaced by the other efficient convention (cf. Fig. 2.3). An approximation of the marginal stationary distribution over aspirations for scenario S0 is depicted in Fig. 2.4, corresponding to the following approximate marginal distribution shown in Tab. 2.1.[11]

Aspirations are concentrated at levels corresponding to the efficient conventions. The asymmetric perfect equilibrium (L, Y) is not favoured over the 'less rational' strategy combinations $(H, \cdot)$. That the Nash equilibrium (H, N)

---

[8] The implicit discrete approximation of the continuous state space E resulting from the finite-byte approximation of real numbers is unproblematic because Φ^η is an open set irreducible T-chain (cf. Sect. 2.6). Linear algebra methods have been applied for control purposes, and Fig. 2.4 is actually based on an explicit 31×21 grid approximation of E and the left-eigenvector of a sampled transition matrix.

[9] This implies that case iii) of Theorem 2.2 applies. Uniform distributions with reasonably wide supports produce qualitatively the same observations.

[10] This high lower bound on inertia has been chosen for a better visualization of the comparative statics. The sensitivity of time shares is notable smaller in absolute terms e.g. for $\tilde{p} = (0.3, 0.3)$ (cf. Fig. 2.5).

[11] The approximation is obtained by a long-run simulation of 20m periods. The initial state has only negligible influence.

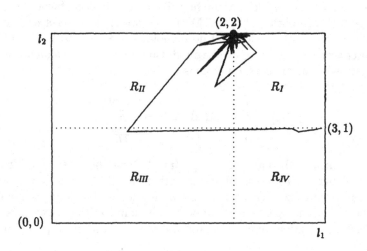

Fig. 2.3. Typical aspiration movements as convention $(H, 2, \cdot, 2)$ is challenged, and eventually replaced by $(L, 3, Y, 1)$

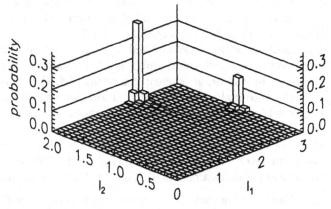

Fig. 2.4. Approximation of the marginal stationary distribution over aspirations in scenario S0

has more than double the frequency of non-Nash equilibrium (H, Y) is not robust to parameter variations. However, robustly even for η and λ quite distant from 0 and 1, respectively, (L, N)-observations are rare in comparison. Therefore, only the frequency of (L, Y)-play is depicted in the following sensitivity analysis; the frequency of $(H, \cdot)$-play is always slightly less than $1 - \text{Prob}(L, Y)$.

Table 2.1. Distribution over action pairs

	LY	*LN*	*HY*	*HN*
Prob(s_1, s_2)	0.367	0.015	0.185	0.433

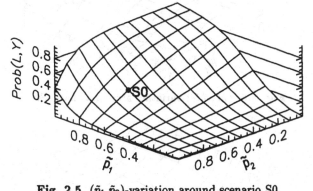

Fig. 2.5. $(\tilde{p}_1, \tilde{p}_2)$-variation around scenario S0

First, $\tilde{p}_1$ and $\tilde{p}_2$ are varied. Given above choice of M_i, inertia $p_i(\Delta)$ is equal to $\tilde{p}_i$ for $i = 1, 2$ when (L, N) is played and aspirations are within the rectangle $R_I = (2, 3) \times (1, 2) \subset L_1 \times L_2$ (see Fig. 2.3). This northeastern part of aspiration space defines a *conflict region* of high aspirations on both sides which cannot be simultaneously satisfied. Adaptation behaviour in the conflict region is crucial for average bargaining outcomes, and players' inertia has the effect of stamina after (L, N)-play: whoever loses patience and switches his action first strongly increases the chances of eventual convergence to the less-favoured (but efficient) convention. With the caveat that Fig. 2.5 gives only *ceteris paribus* information,[12] i.e. all parameters except $\tilde{p}_1$ and $\tilde{p}_2$ are as in scenario S0, one can state:

Observation 2.1. *Player i's average bargaining share increases with his minimal level of action inertia, $\tilde{p}_i$.*

Loosely speaking, it pays to be stubborn or persistent after *major* dissatisfaction. The second varied parameter, slope M_i, defines how drastic player i's response to *minor* dissatisfaction is. This is particularly relevant when e. g. l_1^t is slightly above 2 and (H, L) or (H, N) has been played. One may interpret M_i as a parameter representing a player's irritability. Figure 2.6 then indicates that from a boundedly-rational bargaining perspective it is (weakly) beneficial to be more irritable. More formally stated (with the caveat above):

---

[12] For each $(\tilde{p}_1, \tilde{p}_2)$-combination, Prob$(L, Y)$ has been approximated by 10m periods of Monte-Carlo simulation. The same holds for the numbers depicted in figures 2.6–2.9. *Both* players' average share decreases due to more (L, N)-play when both $\tilde{p}_1$ and $\tilde{p}_2$ are increased. The same holds for simultaneous increases of η_1 and η_2.

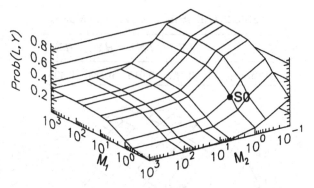

Fig. 2.6. (M_1,M_2)-variation around scenario S0

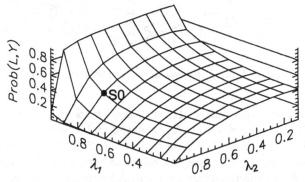

Fig. 2.7. (λ_1,λ_2)-variation around scenario S0

Observation 2.2. *Player i's average bargaining share increases with the slope M_i of her inertia function.*

So, although it pays to be stubborn in response to major dissatisfaction, it is beneficial to be comparatively quickly agitated by minor frustration. Next, aspiration persistence parameter λ_i is varied. Figure 2.7 is summarized (with the above caveat):

Observation 2.3. *Player i's average bargaining share increases with the persistence, λ_i, of his aspirations.*

Again, one may in more colloquial terms infer that it pays to be persistent or stubborn – this time referring to aspiration rather than action updating. The intuition for this is that in the critical conflict region of aspiration space a greater λ_1, for example, decelerates player 1's moves towards 'surrender', i.e. own aspirations below 2, but does not affect moves towards 'victory', i.e. player 2 having aspirations below 1. If one imagines an (unmodeled) encompassing biological or social evolution of agent characters, Observation 2.3 provides some justification for the earlier consideration of the limit case $\lambda \to 1$.

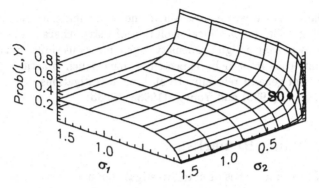

Fig. 2.8. (σ_1,σ_2)-variation around scenario S0

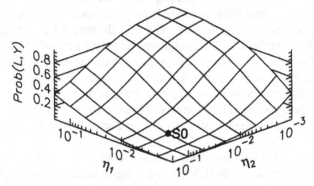

Fig. 2.9. (η_1,η_2)-variation around scenario S0

Finally, figures 2.8 and 2.9 are summarized as follows:

Observation 2.4. *Player i's average bargaining share increases with the standard deviation, σ_i, of her aspiration perturbations.*

Observation 2.5. *Player i's average bargaining share increases with the probability, η_i, of trembles in his aspirations.*

So, a player obtains a bigger average share of the pie when she is capricious, i.e. has frequent and large stochastic variations in her mood or perception of the world. An intuition for this can be found in the asymmetric effect of upward and downward perturbations. An upward tremble in player 1's aspirations when her less-preferred convention $c_{H\cdot}$ is in place results in frustration, which typically will lead to an action switch to L. This bears the chance to establish 1's most-preferred convention c_{LY} for a long time. In contrast, a downward tremble in player 1's aspirations while c_{LY} is in place generally passes by unnoticed and, hence, unexploited by player 2.[13]

---

[13] This suggests that – analogous to the remark on p. 58, fn. 62 – it is beneficial to experience asymmetric trembles.

Observation 2.5 raises a question if one again imagines an unmodeled encompassing social or biological evolution of player characteristics: If it is advantageous to have a high propensity for spontaneous shifts in aspirations, can the limit case $\eta \to 0$ be expected to have much practical relevance? This topic is of general relevance to all limit investigations in the tradition of Kandori et al. (1993) and Young (1993a, 1993b).

2.5 Concluding Remarks

The aim of this chapter has been to investigate boundedly-rational bargaining behaviour in a new framework which, loosely speaking, establishes a 'lowest-rationality' benchmark for bargaining behaviour. An entirely introspective and non-strategic satisficing heuristic based on a gradually adapting endogenous aspiration level has been considered, and the simple bargaining situation represented by the ultimatum minigame has been analysed. The theoretical focus has been on the long-run, asymptotic properties of a Markov chain, in particular the stationary distribution over states which correspond to the different bargaining conventions of perpetual symmetric division, asymmetric division, or disagreement. Simulations of the bargaining and aspiration adaptation process have been studied to obtain both qualitative and numerical results.

Satisficing behaviour has been demonstrated to be 'rational' enough for approximate efficiency in the UMG bargaining situation. However, the resulting distribution of surplus, i. e. the predominance of symmetric or asymmetric division, depends on more specific modelling assumptions, in particular the support of perturbation densities in the efficient conventions. In the simulations, the SPE outcome of the single-shot UMG has not generally been favoured. Varying at most two parameters of the reference scenario at a time, it has turned out beneficial for players to be of persistent character. An advantage of being capricious also has been observed.

The simulations predict that both symmetric and asymmetric surplus-sharing are stable conventions for ultimatum bargaining, and that endogenous switches between them will be observed in the long run. Disagreement is observed with a frequency which increases with the noise level. These results are broadly consistent with observations in bargaining experiments. However, a better fit of the data when the satisficing approach is compared with a purely strategic analysis of the UMG is not surprising in view of the many parameters involved. For proper statements about the practical relevance of above very weak behavioural axioms and their implications, experiments in the vein of Mookherjee and Sopher (1994, 1997), Binmore, McCarthy, Ponti, Samuelson, and Shaked (1999), or Slembeck (1999), and simulations with actual experimental data like Roth and Erev (1995) should be carried out. The property of endogenous transitions between different adaptive equilibria or conventions requires testing in long-run experiments. The – in the author's view – plausible comparative statics observations should also be tested, possibly by having human players interact repeatedly with clearly specified machine players (perhaps, subjects should not know for sure that they play against a computer). If the inherent exploitability of computer players' satisficing behaviour were discovered by the human agents, or the comparative statics contradicted above results, the 'lower bound' on adaptive human bargaining behaviour established in this chapter would need to be raised.

An obvious extension of the model would allow for bigger strategy sets. These could, on the one hand, be used in order to improve the discrete approximation of the original ultimatum game. On the other hand, they would make it possible to analyse a more truly dynamic negotiation between players, e. g. by moving from the ultimatum game to 2-stage or n-stage alternating offers bargaining as investigated in Sect. 1.3.1. It would also be worthwhile to consider two populations of players rather than two fixed agents (cf. the model of Dixon 2000). This could be used to implement a more realistic shock model in which agents' aspirations are also shifted by occasional observations of the success of other agents. Once a population-based model has been built, it is natural to investigate the effects of different interaction structures (see e. g. Ellison 1993, Berninghaus and Schwalbe 1996, or Tieman, Houba, and van der Laan 2000).

The above aspiration adaptation rule could be improved upon. Real players may not immediately experience a fall in aspirations after just one frustrating experience, so that e. g. some lagged or stochastic rule could be more realistic. Most such changes would, however, leave the analytical results in place. Similarly, little would change if agents 'trembled' in implementing their actions. Perturbation of aspirations would then still be needed to ensure Φ^η's ergodicity, and dynamics would remain driven by the four distinct constellations of players' satisfaction and dissatisfaction. An alternative way to motivate a qualitatively similar stochastic process would be to assume random payoffs which approach the deterministic payoffs in Fig. 2.1 as $\eta \to 0$, instead of considering aspiration perturbations. The comparative statics observations would then have different but equally interesting interpretations. For example, it would turn out beneficial to associate quantitatively very volatile consequences with outcomes that have a constant ordinal ranking.

In view of the remarks following Observations 2.3 and 2.5, it could also be worthwhile to explicitly model a higher-level evolution of players' characters. Different average bargaining success could be interpreted as differential fitness in an encompassing imitation or replication process.

Economists' models of human interactive behaviour (and also of simpler decision-making behaviour) can still greatly benefit from collaboration with psychologists. Models, such as the one above, which do without the traditional assumption of optimizing behaviour or the axioms underlying expected utility theory may make it easier to find a common language. The same applies to laboratory experiments. They provide motivation for investigations such as the above, and it would be a success if theoretical progress in capturing human adaptation and learning behaviour can similarly provide motivation for future experimental research.

2.6 Proofs

Proposition 2.2 *Let Φ^0 be defined as above. From any initial state $x_0 \in E$, Φ^0 converges with probability one to a convention $c \in C$.*
Proof. Consider fixed $\lambda \in (0,1)$, an arbitrary initial state $x_0 = (s_1, l_1, s_2, l_2) \in E$, and the corresponding payoffs (π_1, π_2). If for the next t consecutive periods both players stick to their respective action, player 1's aspiration level is $\lambda^t(l_1 - \pi_1) + \pi_1$ according to (2.1).

Infinite repetition of action pair (s_1, s_2) implies convergence to $c_{s_1 s_2}$. The probability for such an infinite run is

$$h(l_1, l_2) := \prod_{t=1}^{\infty} p_1\big(\lambda^t\,(l_1 - \pi_1)\big)\, p_2\big(\lambda^t\,(l_2 - \pi_2)\big),$$

which is positive if and only if

$$\sum_{t=1}^{\infty} \Big[1 - p_1\big(\lambda^t\,(l_1 - \pi_1)\big)\, p_2\big(\lambda^t\,(l_2 - \pi_2)\big)\Big] < \infty. \qquad (2.2)$$

With $K = \max\{K_1, K_2\}$

$$1 - p_1\big(\lambda^t\,(l_1 - \pi_1)\big)\, p_2\big(\lambda^t\,(l_2 - \pi_2)\big)$$
$$\leq 1 - \Big[1 - \big(1 - p_1(\lambda^t\,|l_1 - \pi_1|)\big)\Big]\Big[1 - \big(1 - p_2(\lambda^t\,|l_2 - \pi_2|)\big)\Big]$$
$$\leq 1 - \big[1 - K_1\,\lambda^t\,|l_1 - \pi_1|\big]\big[1 - K_2\,\lambda^t\,|l_2 - \pi_2|\big]$$
$$= 1 - 1 + K_1\,\lambda^t\,|l_1 - \pi_1| + K_2\,\lambda^t\,|l_2 - \pi_2| - K_1\,K_2\,\lambda^{2t}\,|l_1 - \pi_1|\,|l_2 - \pi_2|$$
$$\leq K\,\lambda^t\,\big(|l_1 - \pi_1| + |l_2 - \pi_2|\big),$$

where the first inequality is true because p_i, and thus h, is non-increasing in its argument, and the second inequality follows from $p_i(0) - p_i(\Delta) < K_i\Delta$ with $p_i(0) = 1$ (see p. 86). This implies that the left hand side of (2.2) is bounded above by

$$K\frac{\lambda}{1 - \lambda}\big(|l_1 - \pi_1| + |l_2 - \pi_2|\big),$$

i.e. an infinite run on action pair (s_1, s_2) has positive probability. Because inertia functions are non-increasing in l_1 and l_2, $h(l_1, l_2) \geq h(3, 2) > 0$ for every $(l_1, l_2) \in L_1 \times L_2$, i.e. the probability of an infinite run on the present action pair is bounded away from zero by some $\varepsilon > 0$. The probability of not having started an infinite run on (s_1, s_2) in the last t periods is at most $(1 - \varepsilon)^t$, and converges to zero for $t \to \infty$. $\qquad \square$

Some preliminary definitions and investigations are made before the proofs of Proposition 2.3, Theorem 2.1, and Theorem 2.2. Let $\sigma(E)$ denote the Borel σ-algebra of E. The unperturbed process Φ^0 is formally defined by a *transition probability kernel* $P\colon E \times \sigma(E) \to [0,1]$. $P(x_t, Y)$ is the probability to reach the set $Y \in \sigma(E)$ from state x_t in period $t+1$. Four transitions to singleton sets can have positive probability for Φ^0 – namely those to $x_{t+1} = (s_1, l_1^*(x_t), s_2, l_2^*(x_t))$, with $s_i \in S_i$. For example,

$$P(\underbrace{(L, l_1^t, N, l_2^t)}_{x_t}, \{(H, l_1^*(x_t), N, l_2^*(x_t))\}) = [1 - p_1(\Delta_1(x_t))]\, p_2(\Delta_2(x_t)),$$

where $\Delta_i(x_t)$ is player i's level of disappointment in state x_t. Let the four potential singleton successor sets to x be denoted as $T_j(x)$ with $j \in \{1, \ldots, 4\}$. With $P(x_t, T_j(x_t))$ defined by the action and aspiration update rules of Sect. 2.2, kernel P is given by

$$P(x, Y) = \sum_{j=1}^{4} P(x, T_j(x))\, 1_Y(T_j(x))$$

for all $x \in E$ and $Y \in \sigma(E)$, where 1_Y is the indicator function of set Y.

Consider the partition of aspiration space $L_1 \times L_2$ into $R_I = (2,3) \times (1,2)$, $R_{II} = (0,2) \times (1,2)$, $R_{III} = (0,2) \times (0,1)$, $R_{IV} = (2,3) \times (0,1)$, and the boundaries (see Fig. 2.3). Regions R_{II} to R_{IV} are each characterized by at least one action combination which satisfies both players – stressing players' common interest in coming to some agreement (cf. Fig. 2.10). Dynamics in conflict region R_I are driven by players' opposing interests as to the details of an agreement.

Let Q_1 denote the kernel of the perturbed satisficing process Φ^η conditioned on the event that *only* player 1's aspirations are perturbed:

$$Q_1(x, Y) = P(x, Y_{S_1} \times \{l_1^*(x)\} \times Y_{S_2} \times Y_{L_2}) \int\limits_{l_1 \in Y_{L_1}} g_1(l_1 | l_1^*(x))\, dl_1,$$

for $x \in E$ and $Y \in \sigma(E)$, where Y_{S_i} and Y_{L_i} denote the projection of Y to S_i and L_i, respectively. With similarly defined Q_2 let

$$Q \equiv \frac{1}{2}(Q_1 + Q_2)$$

denote Φ^η's transition rule conditioned on exactly one player experiencing a tremble with both players being equally likely to be the one. Finally,

$$Q_*(x, Y) = P(x, Y_{S_1} \times \{l_1^*(x)\} \times Y_{S_2} \times \{l_2^*(x)\})$$
$$\cdot \int\limits_{l_1 \in Y_{L_1}} \int\limits_{l_2 \in Y_{L_2}} g_1(l_1 | l_1^*(x)) g_2(l_2 | l_2^*(x))\, dl_2 dl_1$$

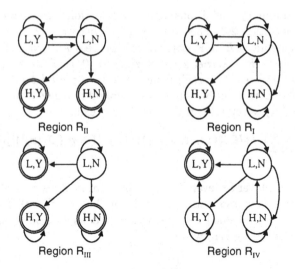

Fig. 2.10. Illustration of action updating in the four regions of aspiration space

is the transition probability kernel conditioned on *both* players' aspirations being perturbed. The complete transition kernel P^η for Φ^η is thus

$$P^\eta \equiv (1 - \eta)^2 P + 2\eta(1 - \eta)Q + \eta^2 Q_*.$$

Q_* is lower semi-continuous (l.s.c.) in x and hence a *strong Feller kernel* (cf. Meyn and Tweedie 1993, p. 128).[14] Since for all $x \in E$ and $Y \in \sigma(E)$

$$P^\eta(x, Y) \geq \eta^2 Q_*(x, Y),$$

$Q_*(x, Y)$ provides a l.s.c. lower bound for P^η. This establishes that Φ^η is a *T-chain* (see Meyn and Tweedie 1993, p. 127), referring to a considerable degree of adaptation of Φ^η to the topology of the state space which is crucial for the convergence results.

Moreover, P^η is *open set irreducible*. This means that every open neighbourhood U_x of every $x \in E$ can be reached from any $y \in E$ in a finite number of transitions: By $L_1 \times L_2$'s compactness and the assumptions on the supports of the post-perturbation densities, g_1 and g_2 define a finite cover of $L_1 \times L_2$ by closed rectangles where moves from one rectangle to a neighbouring one require just one perturbation irrespective of deterministic updating.

---

[14] P is not l.s.c., because it concentrates all weight on $T_j(x)$ with $j \in \{1, \ldots, 4\}$. To see that Q_1 (and hence Q) is not l.s.c. in l_2 (and hence x) consider e. g. set $Y^\varepsilon = \{(s_1, l_1, s_2, l_2^*(x)) \ : \ l_1 \in [l_1^*(x) - \varepsilon, l_1^*(x) + \varepsilon]\}$ for fixed $\varepsilon > 0$, and $l_2 = [l_2^*(x) - (1 - \lambda)\pi_2(s_1, s_2)]/\lambda$. $Q_1((s_1, l_1, s_2, l_2), Y^\varepsilon)$ is strictly positive but one can approach l_2 with some sequence $\{l_2^n \in L_2\}_{n \in \mathbb{N}}$, $l_2^n \to l_2$, such that any $Q_1((s_1, l_1, s_2, l_2^n), Y^\varepsilon)$ is zero.

In particular, any open subset of R_I can thus be reached in finite steps. In R_I, any action profile can be reached with positive probability in at most three steps (cf. Fig. 2.10), and can be preserved by the inertia of players' action choices. After another finite number of transitions the process can thus enter any given open set U. The same argument shows that Q is open set irreducible, too.

Proposition 2.3 *Let the perturbed satisficing process Φ^η be defined as above. For any given perturbation parameter $\eta \in (0,1]$, Φ^η converges (strongly) to a unique limit distribution μ^η which is independent of the initial state $x_0 \in E$.*
Proof. The proposition follows from Theorem 16.2.5 in Meyn and Tweedie (1993, p. 395), stating:

> If Φ^0 is a ψ-irreducible and aperiodic T-chain, and if the state space E is compact, then Φ^0 is uniformly ergodic.

It has just been established that Φ^η is an open set irreducible T-chain. Since this implies ψ-irreducibility (cf. Meyn and Tweedie 1993, p. 133), it only remains to check that Φ^η is aperiodic. In fact, Φ^η is even *strongly aperiodic* (see Meyn and Tweedie 1993, p. 118). First, given the inertia in players' action choices, the supports of perturbation densities g_1 and g_2 define, e. g. for $x^* = c_{HN}$, a neighbourhood U_{x^*} such that $P^\eta(x, U_{x^*}) \geq \nu_1(U_{x^*}) > 0$ for any $x \in U_{x^*}$ and some function ν_1. Second, from any $x \in E$ one can reach U_{x^*} with positive probability in a finite number of steps. $\square$

A few lemmata are considered before Theorems 2.1 and 2.2 are proved. Some of these are quite direct adaptations of results in Karandikar et al. (1998), so only the idea of the proofs is indicated. To start with, define the kernel $R\colon E \times \sigma(E) \to [0,1]$ with

$$R(x,Y) = \lim_{n \to \infty} P^n(x,Y), \qquad x \in E,\ Y \in \sigma(E),$$

where P^n denotes the n-step transition kernel of Φ^0 inductively defined from P. By Proposition 2.2 the above limit exists. R intuitively defines a 'fast-forward' Markov process which moves from x directly to a convention with respective likelihoods for each $c \in C$ defined by Φ^0's long-run behaviour.

Moreover, one can define an artificial cousin, Θ, of the perturbed process Φ^η by considering the result of just *one* perturbation by exactly one player – formally captured by a transition according to Q – and of running the unperturbed process – as concisely described by R – for evermore afterwards. Θ hence has the transition kernel QR with

$$QR\,(x,Y) = Q(x,\cdot)R\,(Y) = \int Q(x,ds)R(s,Y)$$

for $x \in E$ and $Y \in \sigma(E)$.

The long-run behaviour of Φ^η for η close to zero must be similar to that of Θ: For $\eta \to 0$ the average time between two successive perturbations goes to infinity, and hence dynamics of Φ^η become almost identical to that of the unperturbed process. On rare occasions, players' aspirations will experience a tremble with conditional one-step dynamics described by $[2\eta(1-\eta)Q + \eta^2 Q_*]/[1-(1-\eta)^2]$. Since the ratio of simultaneous trembles by both players to single-player trembles, $\eta^2/[2\eta(1-\eta)]$, approaches zero as η vanishes, Q_* will play only a secondary role in defining the long-run average distribution over the state space:

Lemma 2.1. *Let Φ^η be defined as above. The sequence $\{\mu^\eta\}_{\eta \in (0,1]}$ of limit distributions of Φ^η converges weakly to a unique distribution μ^* on $(E, \sigma(E))$ as $\eta \to 0$. μ^* coincides with the unique invariant probability measure of Θ.*

Proof. The result follows directly from Theorem 2 in Karandikar et al. (1998) which states[15]

> Assume that
> [a] For each $x \in E$, $(1/(T+1))\sum_{t=0}^{T} P^t(x, \cdot)$ converges weakly to $R(x, \cdot)$ as $T \to \infty$.
> [b] Q_* has the strong Feller property.
> [c] Q is open set irreducible.
> [d] QR has a unique invariant measure μ^*.
> Then P^η has a unique invariant measure μ^η, which converges weakly to μ^* as $\eta \downarrow 0$.

[a] is an implication of Proposition 2.2. [b] and [c] have already been established to prove Proposition 2.3. It remains to check that QR has a unique invariant measure.

One knows from Proposition 2.1 that any $c \in C$ is an absorbing state for transitions according to R, and from Proposition 2.2 that a transition according to R always results in some $c \in C$ regardless of initial state $x \in E$. Starting in c_{LN}, a transition according to Q leads into R_{III} with positive probability, and from there a transition to any $c \in C$ has positive probability according to R. Starting in $c_{H.}$ or c_{LY}, a transition according to Q leads into region R_I with positive probability, and from there a transition to any $c \in C$ has positive probability according to kernel R. Hence, Θ is ψ-irreducible (cf. Meyn and Tweedie 1993, p. 133), and recurrent (cf. Meyn and Tweedie 1993, p. 182). This implies that Θ has a unique invariant measure (cf. Meyn and Tweedie 1993, Theorem 10.4.9). $\qquad\square$

Next, unperturbed satisficing dynamics Φ^0 in region R_{IV} are investigated. For $\hat{l}_1, \hat{l}_2 \in (0, \frac{1}{2})$, let $I(\hat{l}_1, \hat{l}_2) = [2+\hat{l}_1, 3-\hat{l}_1] \times [\hat{l}_2, 1-\hat{l}_2]$ be a rectangle in R_{IV} (cf. Fig. 2.11).

Lemma 2.2. *Given $\hat{l}_1, \hat{l}_2 \in (0, \frac{1}{2})$ and any $\varepsilon > 0$, there exists $\lambda_1 \in (0,1)$ such that*

---

[15] Karandikar et al. (1998) refer to kernel Q instead of Q_* in [b]. This must be a typo since their kernel Q – just as the above (cf. fn. 14) – is *not* strong Feller, but Q_* is. The latter suffices to establish the T-chain property of P^η which is exploited in their proof.

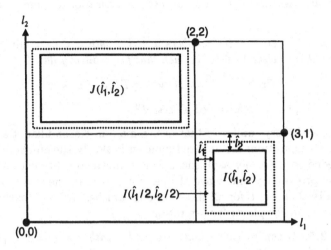

Fig. 2.11. Illustration of $I(\hat{l}_1, \hat{l}_2)$ and $J(\hat{l}_1, \hat{l}_2)$ in Lemmata 2.2 and 2.3

$$^\lambda\text{Prob}(x_t \to c_{LY} \mid x_t = (s_1^t, l_1^t, s_2^t, l_2^t) \wedge (l_1^t, l_2^t) \in I(\hat{l}_1, \hat{l}_2)) > 1 - \varepsilon$$

for all $\lambda \in (\lambda_1, 1)$ and all $t \geq 1$ in unperturbed process Φ^0.

The lemma states that when aspiration updating is slow enough, the players will with arbitrarily high probability settle on the efficient convention c_{LY} when the satisfaction-dissatisfaction constellation is as in R_{IV}, i.e. exactly (L, Y) satisfies both players. The proof is very similar to that of Lemma 2 in Karandikar et al. (1998). One can establish a lower bound on the number of periods that are needed to exit the dotted rectangle $I(\frac{\hat{l}_1}{2}, \frac{\hat{l}_2}{2})$ from anywhere in $I(\hat{l}_1, \hat{l}_2)$. This number goes to infinity as λ approaches 1. Therefore the probability of *not* playing the only mutually satisfying action pair (L, Y) at least once before R_{IV} is left becomes arbitrarily small. However, once (L, Y) is played with aspirations in R_{IV}, Φ^0 converges to c_{LY}.

Considering dynamics in region R_{II}, define rectangles $J(\hat{l}_1, \hat{l}_2) = [\hat{l}_1, 2 - \hat{l}_1] \times [1 + \hat{l}_2, 2 - \hat{l}_2]$ with $\hat{l}_2 \in (0, \frac{1}{2})$ (cf. Fig. 2.11). Analogous to Lemma 2.2 one has:

Lemma 2.3. *Given $\hat{l}_1 \in (0, 1)$, $\hat{l}_2 \in (0, \frac{1}{2})$, and any $\varepsilon > 0$, there exists $\lambda_2 \in (0, 1)$ such that*

$$^\lambda\text{Prob}(x_t \to c_{HY} \vee x_t \to c_{HN} \mid$$
$$x_t = (s_1^t, l_1^t, s_2^t, l_2^t) \wedge (l_1^t, l_2^t) \in J(\hat{l}_1, \hat{l}_2)) > 1 - \varepsilon$$

for all $\lambda \in (\lambda_2, 1)$ and all $t \geq 1$ in unperturbed process Φ^0.

There are two useful results concerning aspiration dynamics in region R_{III}:

Lemma 2.4. *Given positive numbers u and v with $u < w = \min\{2, v\}$ and any $\varepsilon > 0$, there exists $\lambda_3 \in (0, 1)$ such that for arbitrary date T*

$$^\lambda\mathrm{Prob}(l_1^t < u \text{ for some } t \geq T \,|\, l_1^T \geq v) < \varepsilon$$

for all $\lambda \in (\lambda_3, 1)$ in unperturbed process Φ^0.

The proof is very similar to that of Lemma 3 in Karandikar et al. (1998). As player 1's aspirations are updated sufficiently slowly, the number of (L, N)-plays needed to lower her aspirations from v down to u becomes arbitrarily large. The probability that player 1 does not switch to the satisfying H but sticks to dissatisfying L for all these periods vanishes as λ approaches 1. The analogous result for player 2's aspiration is:

Lemma 2.5. *Given positive numbers u and v with $u < w = \min\{1, v\}$ and any $\varepsilon > 0$, there exists $\lambda_4 \in (0, 1)$ such that for arbitrary date T*

$$^\lambda\mathrm{Prob}(l_2^t < u \text{ for some } t \geq T \,|\, l_2^T \geq v) < \varepsilon$$

for all $\lambda \in (\lambda_4, 1)$ in unperturbed process Φ^0.

Lemmata 2.4 and 2.5 put an arbitrarily low upper bound on the probability for aspirations to fall down into region R_{III} from R_I, R_{II}, and R_{IV}. Moreover, conditional on aspirations already lying in R_{III}, a further drop within R_{III} is similarly bounded. Since constant aspirations are impossible outside a convention, R_{III} will effectively be left with arbitrarily high probability when λ is sufficiently close to 1. With these preparations one can establish

Theorem 2.1 *Let the perturbed satisficing process Φ^η be defined as above. The limit stationary distribution of Φ^η for $\eta \to 0$, μ^*, places positive probability only on the Pareto-efficient conventions c_{LY} and $c_{H\cdot}$ as $\lambda \to 1$.*

Proof. By Proposition 2.2, kernel R concentrates all weight on the conventions, and the same must be true for QR. Then, by Lemma 2.1, μ^* must place zero weight on all states except c_{LY}, c_{LN}, and $c_{H\cdot}$.

All aspects of QR relevant to asymptotic behaviour can be captured by a 3×3-matrix $Z = (z_{ij})$ where cell z_{ij} contains the probability of a transition from c_i to c_j ($i, j \in \{LY, LN, H\cdot\}$). This matrix depends on parameter λ.

Given an arbitrary $\lambda < 1$ one can find $\varepsilon(\lambda) > 0$ such that $z_{LY,LN} < \varepsilon(\lambda)$ and $z_{H\cdot,LN} < \varepsilon(\lambda)$ by Lemmata 2.4 and 2.5, and $\varepsilon(\lambda) \to 0$ as $\lambda \to 1$. In contrast, given an arbitrary $\lambda < 1$ there exists $\delta(\lambda) > 0$ such that $z_{LN,LY} + z_{LN,H\cdot} \geq \delta(\lambda)$. This uses that post-perturbation aspiration densities have support in a non-degenerate neighbourhood of c_{LN}. Lemmata 2.4 and 2.5 together with Lemmata 2.2 and 2.3 imply that $\delta(\lambda) \to 1$ as $\lambda \to 1$. This establishes Theorem 2.1.   □

Now consider limit dynamics of Φ^0 in conflict region R_I:

Lemma 2.6. *Given arbitrary* $(l_1, l_2) \in R_I$ *and* $\varepsilon > 0$, *there exists* λ_5 *such that for arbitrary date* $T \geq 1$

$$^\lambda Prob\left(((l_1^t, l_2^t) \in R_{II} \cup R_{IV} \text{ for some } t > T \mid x_T = (s_1^t, l_1, s_2^t, l_2)\right) > 1 - \varepsilon$$

for all $\lambda \in (\lambda_5, 1)$ *in unperturbed satisficing process* Φ^0.

Proof. Consider arbitrary but fixed $(l_1, l_2) \in R_I$. Any direct move from R_I into R_{III} will result in a move back to R_I or to $R_{II} \cup R_{IV}$ with arbitrarily high probability by Lemmata 2.4 and 2.5. The probability of staying in R_I for ever without convergence to a convention is zero by Proposition 2.2. However, the probability of an infinite number of (L, Y)-plays despite dissatisfaction of player 2, which would be necessary to reach c_{LY} from inside R_I, is no more than

$$\prod_{t=1}^{\infty} p_2(\lambda^t(l_2 - 1)).$$

Each factor is decreasing in λ, and the product approaches zero for $\lambda \to 1$. In particular, one can find λ' such that above probability is less than $\varepsilon/2$. Similarly, there is λ' such that the upper bound on the probability of an infinite number of $(H, \cdot)$-plays despite dissatisfaction of player 1 is smaller than $\varepsilon/2$. □

Now, for arbitrary $\lambda_6 \in (0, 1)$ and a fixed aspiration level of player 1 in R_I, $l_1 \in (2, 3)$, one can define

$$\hat{p}_{(L, \cdot)}(l_1) = \sup_{\lambda \in (\lambda_6, 1)} \sup_{l_2 \in (1, 2)} \max_{(s_1, s_2) \in S_1 \times S_2} \; ^\lambda Prob(s_1^{t+2} = L \mid x_t = (s_1, l_1, s_2, l_2))$$

as an upper bound on the probability that player 1 will play L two periods ahead in time when present aspirations are in R_I. By definition, this bound is independent of λ. For $l_1 \in (2, 3)$, $\hat{p}_{(L, \cdot)}(l_1)$ is strictly positive because of player 1's inertia (consider $x_t = (L, l_1, s_2, l_2)$). However, even when an action profile $(L, \cdot)$ has been played in t there is a strictly positive probability for a 2-step transition away from it, namely first to (L, N) and then to $(H, \cdot)$ (cf. Fig. 2.10). This implies $\hat{p}_{(L, \cdot)}(l_1) \leq \hat{p} < 1$ for some $\hat{p} \in (0, 1)$ and all $l_1 \in (2, 3)$.

Similarly, for $l_1 \in (2, 3)$ one can define

$$\check{p}_{(L, N)}(l_1) = \inf_{\lambda \in (\lambda_6, 1)} \inf_{l_2 \in (1, 2)} \min_{(s_1, s_2) \in S_1 \times S_2} \; ^\lambda Prob\left((s_1^{t+2}, s_2^{t+2}) = \right.$$

$$(L, N) \mid x_t = (s_1, l_1, s_2, l_2))$$

as a lower bound on the probability that action pair (L, N) will be played in two periods. Obviously, $\check{p}_{(L, N)}(l_1) \leq \hat{p}_{(L, \cdot)}(l_1)$. Moreover, $\check{p}_{(L, N)}(l_1)$ approaches 0 as $l_1 \downarrow 2$. This is because the probability of player 1 switching away from H, $1 - p_1(l_1 - 2)$, converges to 0. Using the upper bounding assumption on players' inertia functions p_i (cf. Fig. 2.2), there must, however, exist $k > 0$ and $\check{l}_1 > 0$ with $\check{p}_{(L, N)}(l_1) \geq (l_1 - 2)k$ for $l_1 \in (2, 2 + \check{l}_1)$.

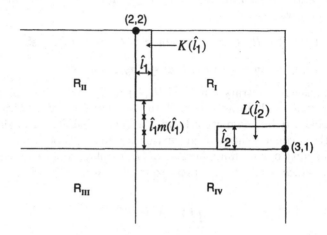

Fig. 2.12. Illustration of $K(\hat{l}_1)$ and $L(\hat{l}_2)$ in Lemmata 2.7 and 2.8

For any $\hat{l}_1 \in (0,1)$, choose

$$m(\hat{l}_1) = \max \left\{ \frac{2 \ln \breve{p}_{(L,N)}(2 + \hat{l}_1)}{\ln \hat{p}_{(L,\cdot)}(2 + \hat{l}_1)} , 3 \right\}.$$

Since $\breve{p}_{(L,N)}(2 + \hat{l}_1)$ goes to zero but $\hat{p}_{(L,\cdot)}(2 + \hat{l}_1)$ is bounded away from both zero and one, $m(\hat{l}_1) \to \infty$ as $\hat{l}_1 \to 0$.

Finally, define rectangles $K(\hat{l}_1) = (2, 2 + \hat{l}_1) \times (1 + \hat{l}_1\, m(\hat{l}_1), 2)$ for all $\hat{l}_1$ such that $\hat{l}_1\, m(\hat{l}_1) < 1$, also referred to as *permissable* $\hat{l}_1$ (cf. Fig. 2.12). Using the bounds on $\breve{p}_{(L,N)}(l_1)$ and $\hat{p}_{(L,\cdot)}(l_1)$ for $l_1 \in (2,3)$, one gets

$$\frac{2 \ln(\hat{l}_1\, k)}{\ln \hat{p}} + 3 \geq m(\hat{l}_1) \geq 3.$$

From this follows $\hat{l}_1\, m(\hat{l}_1) \to 0$ as $\hat{l}_1 \to 0$, ensuring that there is always an $\bar{l}_1$ independent of λ such that all $\hat{l}_1 \in (0, \bar{l}_1)$ are permissable.

Lemma 2.7. *From arbitrary state x_T with $(l_1^T, l_2^T) \in K(\hat{l}_1)$ for an arbitrary permissable $\hat{l}_1$, a move by unperturbed process Φ^0 into R_{II} becomes infinitely more likely as $\lambda \to 1$ than a move into R_{IV}.*

Proof. Consider an arbitrary but fixed state x_T with $(l_1^T, l_2^T) \in K(\hat{l}_1)$ for some fixed permissable $\hat{l}_1$, and let λ_6 in the definitions of $\breve{p}_{(L,N)}$ and $\hat{p}_{(L,\cdot)}$ be large enough such that $\frac{\hat{l}_1}{2(1 - \lambda_6)} \geq 2$. One round of (L, N)-play lowers player 1's aspiration by at least $2(1 - \lambda)$, and player 2's aspiration by at most the amount $2(1 - \lambda)$. For player 2, the aspiration decrease caused by (L, N)-play is greater than that of (L, Y)-play.

The probability of a move to R_{II} is at least as high as that of observing

$$T^*(\hat{l}_1, \lambda) = \left\lceil \frac{\hat{l}_1}{2(1 - \lambda)} \right\rceil$$

periods of consecutive (L, N) play.

$$T^{**}(\hat{l}_1, \lambda) = \left\lceil \frac{\hat{l}_1 \, m(\hat{l}_1)}{2(1 - \lambda)} \right\rceil$$

is the minimal number of $(L, \cdot)$-plays which could decrease player 2's aspiration level below 1 and hence lead aspirations into R_{IV}.

Starting with x_T and considering periods $x_{T+2}, \ldots, x_{T+T^*+2}$, one gets $\check{p}_{(L,N)}(2 + \hat{l}_1)^{T^*(\hat{l}_1, \lambda)}$ as a lower bound on the probability to move into R_{II}. Similarly, considering periods $x_{T+2}, x_{T+3}, \ldots$ one obtains $\hat{p}_{(L,\cdot)}(2+\hat{l}_1)^{T^{**}(\hat{l}_1, \lambda)}$ as an upper bound on the probability of moving to R_{IV}.

The ratio of the probabilities of moving from $K(\hat{l}_1)$ into region R_{II} and R_{IV}, respectively, is

$$
\begin{aligned}
r(\lambda) &= \frac{{}^\lambda \mathrm{Prob}(\,(l_1^t, l_2^t) \in R_{II} \text{ for some } t \geq T \,|\, (l_1^T, l_2^T) \in K(\hat{l}_1)\,)}{{}^\lambda \mathrm{Prob}(\,(l_1^t, l_2^t) \in R_{IV} \text{ for some } t \geq T \,|\, (l_1^T, l_2^T) \in K(\hat{l}_1)\,)} \\[2mm]
&\geq \frac{\check{p}_{(L,N)}(2 + \hat{l}_1)^{T^*(\hat{l}_1, \lambda)}}{\hat{p}_{(L,\cdot)}(2 + \hat{l}_1)^{T^{**}(\hat{l}_1, \lambda)}} \\[2mm]
&= \frac{\check{p}_{(L,N)}(2 + \hat{l}_1)^{\lceil \hat{l}_1 / [2(1-\lambda)] \rceil}}{\hat{p}_{(L,\cdot)}(2 + \hat{l}_1)^{\lceil \hat{l}_1 \, m(\hat{l}_1) / [2(1-\lambda)] \rceil}} \\[2mm]
&\geq \frac{\check{p}_{(L,N)}(2 + \hat{l}_1)^{2\hat{l}_1 / [2(1-\lambda)]}}{\hat{p}_{(L,\cdot)}(2 + \hat{l}_1)^{m(\hat{l}_1)\hat{l}_1 / [2(1-\lambda)]}} \\[2mm]
&\geq \left(\frac{\check{p}_{(L,N)}(2 + \hat{l}_1)^2}{\hat{p}_{(L,\cdot)}(2 + \hat{l}_1)^{m(\hat{l}_1)}} \right)^{\hat{l}_1 / [2(1-\lambda)]} .
\end{aligned}
$$

Given the choice of $m(\hat{l}_1)$, the base term is greater than 1. The exponent goes to infinity as λ approaches 1, implying $r(\lambda) \xrightarrow[\lambda \to \infty]{} \infty$. $\qquad\square$

With analogously defined rectangles $L(\hat{l}_2)$ (cf. Fig. 2.12) and analogous permissibility condition, one has:

Lemma 2.8. *From arbitrary state x_T with $(l_1^T, l_2^T) \in L(\hat{l}_2)$ for an arbitrary permissable $\hat{l}_2$, a move by unperturbed process Φ^0 into R_{IV} becomes infinitely more likely as $\lambda \to 1$ than a move into R_{II}.*

Lemmata 2.7 and 2.8 stress that, even for $\lambda \to 1$, the unperturbed satisficing process Φ^0 does not single out a particular bargaining convention. Rather, a start in $L(\hat{l}_2)$ will lead to R_{IV} by Lemma 2.8 and result in c_{LY} by Lemma 2.2; an initial state in $K(\hat{l}_1)$ will by Lemmata 2.7 and 2.3 result in $c_{H\cdot}$. With this, one can finally prove

Theorem 2.2 *Let the perturbed satisficing process Φ^η be defined as above, and let μ^* denote the limit stationary distribution of Φ^η for $\eta \to 0$. The supports of perturbation densities in conventions $c_{H\cdot}$ and c_{LY} can be chosen*

  i) *such that μ^* places all weight on the asymmetric efficient convention c_{LY},*
  ii) *such that μ^* places all weight on the symmetric efficient convention $c_{H\cdot}$,*
      or
iii) *such that μ^* places positive weight on both c_{LY} and $c_{H\cdot}$,*

as $\lambda \to 1$. In particular, when post-perturbation aspiration densities have full support in $L_1 \times L_2$, then μ^* places positive weight on both c_{LY} and $c_{H\cdot}$.
Proof. Let z_{jk} refer to elements of the matrix defined in the proof of Theorem 2.1, where $z_{LY,LN}$ and $z_{H\cdot,LN}$ approach 0 for $\lambda \to 1$. The assumptions on g_1 and g_2 ensure that a positive measure of perturbations from c_{LY} or $c_{H\cdot}$ stays in a neighbourhood comprised in $L(\hat{l}_2)$ or $K(\hat{l}_1)$ respectively for permissable $\hat{l}_2$ and $\hat{l}_1$ (cf. Lemmata 2.7 and 2.8), implying an almost sure return to c_{LY} or $c_{H\cdot}$, respectively, as $\lambda \to 1$. So, $z_{LY,LY}, z_{H\cdot,H\cdot} \geq \nu > 0$.
  i) Choose the supports of $g_i(\cdot|c_{H\cdot})$ $(i = 1, 2)$ wide enough to place positive weight on some rectangle $L(\hat{l}_2)$ for a permissable $\hat{l}_2$, but choose $g_i(\cdot|c_{LY})$ such that they place all weight on some rectangle $L(\hat{l}'_2)$ for a permissable $\hat{l}'_2$ and on R_{IV}. Then, $z_{LY,H\cdot} < \varepsilon(\lambda)$ but $z_{H\cdot,LY} \geq \delta > 0$. As $\lambda \to 1$, $\varepsilon(\lambda) \to 0$ by Lemmata 2.8 and 2.2. In the limit c_{LY} is the only absorbing state of Φ^η, and μ^* places all weight on it.
  ii) Choose the supports of $g_i(\cdot|c_{H\cdot})$ $(i = 1, 2)$ such that they place all weight on some rectangle $K(\hat{l}_1)$ for permissable $\hat{l}_1$ and R_{II}, but choose those of $g_i(\cdot|c_{LY})$ wide enough to place positive weight on some rectangle $K(\hat{l}'_1)$ for a permissable $\hat{l}'_1$. This time, $z_{LY,H\cdot} \geq \delta > 0$ but $z_{H\cdot,LY} < \varepsilon(\lambda)$ and $\varepsilon(\lambda) \to 0$ for $\lambda \to 1$ by Lemmata 2.7 and 2.3. In the limit, $c_{H\cdot}$ becomes the only absorbing state, and μ^* places all weight on it.
  iii) Choose the supports of $g_i(\cdot|c_{H\cdot})$ $(i = 1, 2)$ wide enough to place positive weight on some rectangle $L(\hat{l}_2)$ for a permissable $\hat{l}_2$, and choose those of $g_i(\cdot|c_{LY})$ wide enough to place positive weight on some rectangle $K(\hat{l}_1)$ for a permissable $\hat{l}_1$. A special case is that post-perturbation aspiration densities have full support in $L_1 \times L_2$. Now, both $z_{LY,H\cdot}$ and $z_{H\cdot,LY}$ are bounded away from zero. Consequently, μ^* will place positive weight on both c_{LY} and $c_{H\cdot}$ even in the limit $\lambda \to 1$. $\square$

3. Bilateral Bargaining and Decision Power

Simple games are n-player cooperative games in which each subset of players can be classified as either a winning coalition or a losing coalition. They can be used to model economic or political decision bodies like parliaments or shareholder meetings in which proposals are either passed or rejected. *Power indices* are functions that map n-person simple games to n-dimensional real vectors. They assign to each player a number that indicates the player's a priori power to shape events, and they have been applied to evaluate numerous political and economic institutions in practice. Power distributions in the context of shareholders' meetings have been one focus of attention, with the theoretical challenge of accounting for cross-ownership (see e. g. Leech 1988 and Gambarelli and Owen 1994). In the political sphere, decision making in the U. S. Congress, U. S. presidential elections, the U. N. Security Council, and the institutions of the European Union have all been studied extensively using power indices.[1] Power measurement techniques have played an important role in the discussion of institutional amendments in preparation of an enlargement of the European Union at the Nice 2000 summit (see e. g. The Economist, November 25th 2000, p. 126).

Despite the wide application and almost fifty years after the seminal contribution to power measurement by Shapley and Shubik (1954), there is still considerable controversy as to what constitutes an appropriate power measure. In the wake of Shapley and Shubik's work, numerous power indices have been proposed – most notably by Banzhaf (1965), Deegan and Packel (1978), and Holler and Packel (1983).[2] These power indices satisfy a number of requirements or axioms. For example, the power of a dictator, who forms a winning coalition himself, is scaled to unity; that of null or dummy players, who cannot turn any losing coalition into a winning one by joining, is set to zero. Anonymity, particular monotonicity and aggregation properties for different simple games are additionally required or, at least, desired from a reasonable index. However, strategic aspects of bargaining among players have generally been neglected. This explains that none of the established in-

---

[1] Compare e. g. Owen (1975), Kirman and Widgrén (1995), Laruelle and Widgrén (1998), and Nurmi (1998, ch. 7).

[2] Straffin (1994) gives a worthwhile overview. See Felsenthal and Machover (1998) for a detailed comparative investigation of power indices, their properties, and applicability.

dices is consistent with traditional notions of competitive equilibrium or the cooperative concept of the core; a substantial amount of power is indicated for players who according to all other approaches are powerless.

The aim of this chapter is to apply elements of bilateral bargaining theory to the measurement of power in order to overcome this deficiency. Namely, the concept of *inferior players* is developed based on bilateral ultimatum bargaining, and thereafter applied in the traditional axiomatic and probabilistic frameworks of power measurement. It is argued that the dummy axiom conventionally used in power measurement should be replaced by a stricter axiom based on this new concept. The proposed axiom requires indices to *not* take into account a player's supposed power (as traditionally measured by swings, pivot positions, etc.) if some other player can credibly issue the following ultimatum to him: Accept (almost) no share of the spoils from a winning coalition or be prevented from taking part in one at all. Thus, power measurement is brought more in line with competitive and non-cooperative analysis.

Section 3.1 introduces principal elements of the measurement of decision power. Section 3.2 develops the concept of inferior players and proposes the inferior player axiom. The strict power index (SPI), related to the Banzhaf index, is introduced and axiomatized in Sect. 3.3. Then, Sect. 3.4 investigates inferiority in the realm of probabilistic power measurement. A probabilistic foundation of the SPI and a more general family of indices is given, before Sect. 3.5 remarks on promising extensions of the new concepts presented in this chapter. The presentation draws heavily on Napel and Widgrén (2000) and Napel and Widgrén (2001b).[3]

3.1 Power Indices

Let $I = \{1, 2, \ldots, n\}$ be the set of players. $\wp(I) = \{0, 1\}^n$ is the set of feasible coalitions. A *(monotonic) simple game* v is characterized by the set $W(v) \subsetneq \wp(I)$ of *winning coalitions*. $W(v)$ satisfies $\varnothing \notin W(v)$, $I \in W(v)$ and $S \in W(v) \wedge S \subseteq T \Rightarrow T \in W(v)$. Using the last property, a concise description of v can be given by

$$M(v) := \{S \subseteq I : S \in W(v) \wedge (\forall i \in S) : S \setminus \{i\} \notin W(v)\},$$

the set of *minimal winning coalitions (MWC)*. Game v can be identified with the *characteristic function* $v : \wp(I) \to \{0, 1\}$ defined by

$$v(S) := \begin{cases} 0; & S \notin W(v), \\ 1; & S \in W(v). \end{cases}$$

---

[3] See also Napel and Widgrén (2001a).

Let $\mathcal{G}^I$ denote the set of all such n-person simple games. *Weighted voting games* are special instances of simple games that are characterized by a non-negative real vector $r_v = (q; w_1, \ldots, w_n)$, where w_i represents player i's voting weight and q represents the quota of votes that establishes a winning coalition.

A player who by leaving a winning coalition $S \in W(v)$ turns it into a losing coalition $S \setminus \{i\} \notin W(v)$ has a *swing* in S, and is called a *crucial member* of coalition S. Coalitions where player i has a swing are called *crucial coalitions with respect to i*. Let

$$C_i(v) := \{S \subseteq I : S \in W(v) \wedge S \setminus \{i\} \notin W(v)\}$$

denote the set of crucial coalitions w.r.t. i. The number of swings of player i will be denoted by

$$\eta_i(v) := |C_i(v)|.$$

The total number of swings in a simple game v is $\eta_\Sigma(v) = \sum_{i=1}^n \eta_i(v)$. A player i with $\eta_i(v) = 0$ is called a *dummy player*.

A power index is a mapping $\mu \colon \mathcal{G}^I \to \mathbb{R}_+^n$, assigning to each player $i \in I$ a number $\mu_i(v)$ that indicates i's a priori power in the considered game v. Several properties are typically required from a power index. For example, all established indices indicate zero power for dummy players, i.e. they satisfy the following *dummy player axiom*:

Dummy players (DUM):
$(\forall v \in \mathcal{G}^I) \colon \{\eta_i(v) = 0 \implies \mu_i(v) = 0\}$.

Equally natural is the following *symmetry* or *anonymity axiom*:[4]

Anonymity (ANY):
For every simple game $v \in \mathcal{G}^I$ and every permutation ϱ of I: $\mu_{\varrho(i)}(\varrho v) = \mu_i(v)$.

Here, ϱv is defined by $(\varrho v)(S) := v(\varrho^{-1}(S))$. Typically, one scales an index μ such that $\mu_i(v) = 1$ if and only if i is a dictator in v, i.e. $M(v) = \{\{i\}\}$. Two alternative axioms are commonly introduced to ensure this together with (DUM):

Absolute power (ABS):
$(\forall v \in \mathcal{G}^I) \colon \sum_{i=1}^n \mu_i(v) = \frac{\eta_\Sigma(v)}{2^{n-1}}$.

Relative power (REL):
$(\forall v \in \mathcal{G}^I) \colon \sum_{i=1}^n \mu_i(v) = 1$.

Given $u, v \in \mathcal{G}^I$, let the simple game $u \vee v \in \mathcal{G}^I$ be defined by the characteristic function $(u \vee v)(S) := \max\{u(S), v(S)\}$ for all $S \subseteq I$. Similarly, define $u \wedge v \in \mathcal{G}^I$ by $(u \wedge v)(S) := \min\{u(S), v(S)\}$. It is often useful that any simple game

---

[4] *Weighted values* serve as an example where anonymity is relaxed (see Kalai and Samet 1987 for details).

$v \in \mathcal{G}^I$ can be written as the composition $u_{S_1} \vee \ldots \vee u_{S_r}$, where $M(v) = \{S_1, \ldots, S_r\}$ and u_{S_k} is the **unanimity** or **auxiliary game** in which exactly all coalitions containing v's MWC S_k are winning. Some power measures are based on a linear notion of power which explicitly requires from a power index μ that the *additivity axiom* holds:

Additivity (ADD):

$(\forall u, v \in \mathcal{G}^I)\colon \mu(u \vee v) = \mu(u) + \mu(v) - \mu(u \wedge v).$

For the class of weighted voting games it is an intuitive requirement that more power is indicated for players with more voting weight:

Local monotonicity (LOC):

If $v \in \mathcal{G}^I$ has the weighted voting game representation $r_v = (q; w_1, \ldots, w_n)$, then $w_i \geq w_j \implies \mu_i(v) \geq \mu_j(v)$.

Monotonicity can also be defined with respect to players' positions in different simple games (cf. e. g. Levínský and Silárszky 2001). $u \in \mathcal{G}^I$ can be considered 'better' than $v \in \mathcal{G}^I$ from player i's point of view if all winning coalitions of v with i also win in u (and, possibly, some other coalitions with i win in u) and if all winning coalitions of u without i also win in v (and possibly some more). Formally, define the partial orderings $>_i$ with

$$u >_i v \;:\Longleftrightarrow\; \begin{cases} i \in S \wedge S \in W(v) \Rightarrow S \in W(u) \\ \wedge\, i \notin S \wedge S \in W(u) \Rightarrow S \in W(v) \end{cases}$$

for $i \in I$. With this, one can require

Global monotonicity (GLO):

$(\forall u, v \in \mathcal{G}^I)\colon \{u >_i v \implies \mu_i(u) \geq \mu_i(v)\}$

Provided that an index is anonymous, (GLO) implies (LOC) (Levínský and Silárszky 2001).

The oldest power index is usually named *Banzhaf index (BZI)* after its definition by Banzhaf (1965), although one of its several variants has earlier been proposed by Penrose (1946).[5] The BZI $\beta\colon \mathcal{G}^I \to \mathbb{R}^n_+$ is defined by

$$\beta_i(v) := \sum_{\substack{S \subseteq I \\ i \in S}} \frac{v(S) - v(S \setminus \{i\})}{2^{n-1}} = \sum_{S \in C_i(v)} \frac{1}{2^{n-1}} = \frac{\eta_i(v)}{2^{n-1}}.$$

The BZI gives equal weight to every swing in the game. Since there are 2^{n-1} coalitions in which player i could have a swing, $\beta_i(v)$ represents i's ratio of actual to potential number of swings. The BZI can be axiomatically characterized as the unique index which satisfies (DUM), (ANY), (ABS), and

---

[5] Cf. Felsenthal and Machover (1998, pp. 6ff) for a historical account. Riker (1986) even traces back the intuition underlying the BZI to arguments put forward by a Maryland delegate, called Luther Martin, to the U. S. Constitutional Convention in 1787. He interprets this as showing "that power indices are not merely mathematicians' fancies but obvious categories of thought for practical politicians" (p. 294).

(ADD) (Dubey and Shapley 1979).[6] It satisfies (GLO) and hence (LOC). A normalized version of the BZI is frequently used in applications. Normalization replaces (ABS) with (REL) but violates (ADD) and (GLO); the ratio interpretation is lost as well as the BZI's probabilistic foundation given below.

Another prominent power index has been proposed by Shapley and Shubik (1954) based on the Shapley value (Shapley 1953). The *Shapley-Shubik index* *(SSI)* $\varphi \colon \mathcal{G}^I \to \mathbb{R}^n_+$ is defined by

$$\varphi_i(v) := \sum_{\substack{S \subseteq I \\ i \in S}} \frac{(|S| - 1)! \, (n - |S|)!}{n!} \Big[v(S) - v(S \setminus \{i\}) \Big]$$

$$= \sum_{S \in C_i(v)} \frac{(|S| - 1)! \, (n - |S|)!}{n!}.$$

Considering a fixed player i, both SSI and BZI are a weighted average of i's *marginal contribution* $v(S) - v(S \setminus \{i\})$ to all coalitions $S \subseteq I$. The SSI weights player i's swing in coalition S with the number of orderings $(i_1, \ldots, i_{|S|}, \ldots, i_n)$ of all players $j \in I$ such that $\{i_1, \ldots, i_{|S|}\} = S$ and $i_{|S|} = i$. Given $(i_1, \ldots, i_{|S|}, \ldots, i_n)$, player $i_{|S|}$ is said to have a *pivot position* in this ordering if and only if $v(\{i_1, \ldots, i_{|S|}\}) = 1$ and $v(\{i_1, \ldots, i_{|S|-1}\}) = 0$. Thus, the SSI gives equal weight to every pivot position in the game rather than every swing. The SSI is the unique index satisfying (DUM), (ANY), (REL), and (ADD). It also satisfies (GLO).

The *Deegan-Packel index* (introduced and axiomatically characterized by Deegan and Packel 1978) only considers minimal winning coalitions, and weights each swing of player i in a coalition $S \in M(v)$ with $1/|S|$, i.e. it distributes imagined spoils of forming a MWC equally among its members. The resulting numbers are then normalized to satisfy (REL). The *Holler-Packel index* or *public good index* (introduced by Holler 1978, 1982a, and axiomatized by Holler and Packel 1983 and Napel 1999b, 2001) also considers only swings in MWC, but distributes imagined spoils like a public good or a club good – i.e. each such swing receives weight 1 – before normalization. Neither Deegan-Packel index nor Holler-Packel index satisfy (LOC).[7]

Power in simple games can also be analysed in a probabilistic setting. Instead of deterministic coalitions $S \subseteq I$, corresponding to corner points $s \in \{0, 1\}^n$ of the n-dimensional unit cube, one considers fuzzy or random coalitions $\mathfrak{S}$ represented by points $p \in [0, 1]^n$ anywhere in the cube. Each $p_i \in [0, 1]$ is interpreted as the probability of player $i \in I$ deciding in favour of a random proposal or of participating in a randomly formed coalition; it is referred to as player i's *acceptance rate*.

---

[6] Laruelle and Valenciano (2001) provide a different, perhaps more natural axiomatization for the BZI.

[7] That this is not necessarily a cause for concern is argued, for example, by Holler (1997).

Players' acceptance decisions are assumed to be independent. Thus, the probability of forming a given coalition $S \subseteq I$ is $\text{Prob}(\mathfrak{S} = S) = \Pi_{i \in S} p_i \Pi_{j \notin S}(1 - p_j)$. The characteristic function $v \colon \{0,1\}^n \to \{0,1\}$ of a simple game can be extended by weighting $v(S)$ for all coalitions $S \subseteq I$ with their respective probability of formation. One obtains the *multilinear extension (MLE)* $f \colon [0,1]^n \to [0,1]$ of game v (see Owen 1972, 1988):

$$f(p_1, \ldots, p_n) := \sum_{S \subseteq I} \prod_{i \in S} p_i \prod_{j \notin S} (1 - p_j)\, v\,(S)$$
$$= \sum_{S \in W(v)} \prod_{i \in S} p_i \prod_{j \notin S} (1 - p_j)\,.$$

For fixed acceptance rates $(p_1, \ldots, p_n)$, the MLE gives the probability of formation of a winning coalition in v, and also the expected value of v. Note that players' acceptance rates may not be constants, but random variables themselves.

Let f_i denote the partial derivative $\partial f / \partial p_i$ of v's MLE with respect to p_i. It is usually referred to as player i's *power polynomial* (Straffin 1977, 1988). $f_i(p_1, \ldots, p_n)$ is the probability of i having a swing in the random coalition to be formed in game v. When players' acceptance rates $(p_1, \ldots, p_n)$ are random variables with a joint distribution P, the expectation

$$Ef_i = \int f_i\,(p_1, \ldots, p_n)\, dP \tag{3.1}$$

is an indicator of i's power in game v. The *probabilistic power index* defined by (3.1) coincides with the traditional deterministic formulation of power indices for several plausible probability models. In particular, when all players' acceptance rates are independently drawn from a uniform distribution on $[0,1]$ – in short notation: $(\forall i \in I) \colon p_i \overset{i.i.d.}{\sim} U[0,1]$ – then (3.1) equals the BZI. The SSI is obtained from the more restrictive assumption that t is uniformly distributed on $[0,1]$ and $(\forall i \in I) \colon p_i = t$.

The probabilistic representation of power indices highlights a close link to measures of *structural importance* and *reliability importance* in engineering (see, for example, Barlow and Proschan 1975). There, the BZI is also known as the *Birnbaum index*, after its definition by Birnbaum (1969). It measures the importance of individual technical components that work in parallel or in series with other electrical or mechanical devices in multi-component systems.[8] In the light of the fundamental differences between technical systems and political or economic institutions, this great range of applications is flattering for the BZI, but also quite surprising. Should one not expect real players' ability to threaten, blackmail, and outguess one another to make a difference?

---

[8] The *structure function* v indicates whether the system constructed from $I = \{1, 2, \ldots, n\}$ components works given a particular constellation $s \in \{0,1\}^n$ of functioning ($s_i = 1$) and defect components ($s_i = 0$). The multilinear extension (3.1) is also known as the *reliability function*, where p_i is component i's probability of functioning.

3.2 Inferior Players

Consider a federal government A that needs approval from at least one of two provincial governments, B and C, to pass laws. Alternatively, let A be a shareholder who needs to be backed by at least one of two smaller shareholders to decide on questions of corporate policy. Both situations can abstractly be modelled as the 3-player simple game v_1 with $I = \{A, B, C\}$ and $M(v_1) = \{AB, AC\}$.[9] Intuitive economic analysis would claim A to be 'on the short side of the market,' implying that B and C cannot influence terms of trade. From the perspective of non-cooperative bilateral bargaining theory, player A can credibly issue an ultimatum to B (or C) in which A proposes to establish coalition AB in return for (in the limit) total concession by B on those economic or policy issues related to the formation of a winning coalition on which A and B have opposing interests. Thus, B is robbed of the power commonly associated with his swing. The possibility of A flipping a coin before B or C is chosen to establish a winning coalition extends the argument to both players. Drawing on cooperative game theory,[10] the core and the nucleolus of this game are $\{(1, 0, 0)\}$ and support the intuition that B and C are powerless.

Despite these consistent results of different types of reasoning, the power indices of Banzhaf, Shapley-Shubik, Deegan-Packel, or Holler-Packel indicate substantial power for powerless players B and C in this simple reference game. They yield the normalized vectors $(\frac{3}{5}, \frac{1}{5}, \frac{1}{5})$, $(\frac{2}{3}, \frac{1}{6}, \frac{1}{6})$, $(\frac{1}{2}, \frac{1}{4}, \frac{1}{4})$, and $(\frac{1}{2}, \frac{1}{4}, \frac{1}{4})$, respectively. The divergence between these power indications based on conventional indices on the one hand, and arguments based on bilateral bargaining theory, competitive analysis, and the core or nucleolus on the other hand is motivation to formally capture the sense in which player B is 'weak' in v_1, and to correspondingly modify established indices.

Describing B's position a bit more abstractly, it can be said that there exists a player who can veto all coalitions in which B makes a positive contribution, i. e. is crucial, but who can herself form a crucial coalition without an opportunity for B to interfere. Threatened by A taking this outside option, B prefers (almost) any concession to A's demands to being excluded from winning; A can credibly initiate an ultimatum game (see Sect. 1.3.1) that has B in the role of the responder. In this sense, B is an *inferior player* in game v_1. Formalizing this intuitive notion of inferiority, one can state:

Definition 3.1. *Player i is* inferior *in simple game v if* $(\exists j \neq i)$:

$$(\forall S \in C_i(v)): \quad j \in S$$
$$\wedge \ (\exists S' \in C_j(v)): \quad i \notin S'$$

---

[9] For simpler exposition, AB is used to denote $\{A, B\}$ or, equivalently, the corner point $(1, 1, 0)$ of the unit cube.

[10] Owen (1995) gives an excellent introduction to the core, nucleolus, and other cooperative solution concepts.

Let $I^*(v) \subsetneq I$ denote the set of inferior players in v. There is a neat equivalent definition:

Proposition 3.1. *Player i is inferior in $v \in \mathcal{G}^I$* $\iff$ *$(\exists j \neq i)$: $C_i(v) \subsetneq C_j(v)$.*

Proof. a) Let i be inferior in v. Assume that there exists $\tilde{S} \in C_i(v)$ with $\tilde{S} \notin C_j(v)$. It follows that $\tilde{S} \in W(v)$, and $\tilde{S} \setminus \{j\} \in W(v)$. Furthermore, from $\tilde{S} \setminus \{i\} \notin W(v)$ it follows that $\tilde{S} \setminus \{j\} \setminus \{i\} \notin W(v)$. Thus, $\tilde{S} \setminus \{j\} \in C_i(v)$ – a contradiction to $(\forall S \in C_i(v))$: $j \in S$. So $C_i(v) \subseteq C_j(v)$. Because j is crucial in at least one coalition S' without i, $C_i(v) \subsetneq C_j(v)$.

b) $S \in C_j(v)$ implies $j \in S$ – establishing the first part of Definition 3.1. Assume $C_i(v) \subsetneq C_j(v)$ and $(\forall S' \in C_j(v))$: $i \in S'$. Using the argument in a), the latter implies $C_j(v) \subseteq C_i(v)$. This is a contradiction. $\qquad\square$

Any dummy player is inferior. The reverse is true for *strong* or *decisive simple games* where $(\forall S \subseteq I)$: $\{S \in W(v) \vee I \setminus S \in W(v)\}$. In such games, a swing for player i in coalition S implies a swing also in coalition with the disjoint set of partners $N \setminus S$; thereby every swing truly means power.

Existence of inferior players does not require a veto player such as A in v_1. An example for this is the simple game v_2 with $I = \{A, B, C, D, E\}$ and $M(v_2) = \{ABC, ABD, ABE, ACD, BCDE\}$: There is no veto player, but player E is crucial only in coalitions with B and hence inferior. In contrast, player A is crucial without B and E (in ACD), C (in ABD), and D (in ABE), and therefore is no inferior player. It can similarly be verified that players B, C, and E, are crucial at least once without any other player, i.e. they are not inferior either. This illustrates that inferiority does not rest on the perfect substitutability of two players, such as B and C in v_1.

A player i can be inferior because of a player j who is himself inferior. However, by Proposition 3.1 and the transitivity of $\subsetneq$, there is at least one non-inferior player k who makes i inferior. If some player i is inferior in v, $x_i = 0$ for any element x of v's core. Rotating members of the U.N. Security Council are a real-world example of inferior players.

The concept of inferior players is a simple way to classify players. It defines a partition, not a partial ordering on I. It is not based on any ordering of players either, e.g. the desirability relation of Maschler and Peleg (1966). With the latter, a classification of non-dummy players into *sum* and *step* players has been defined by Ostmann (1987). A player i who is strictly less desirable than a step j is shown to be "ruled" by the latter in the same sense as an inferior player, i.e. not having swings without j while j has some without i. The concept of inferior players requires a considerably smaller theoretical apparatus than that of steps and their followers. The calculation of the partial ordering induced by individual desirability is often quite complex; then a series of further calculations is required. In contrast, it is very directly checked whether a player is inferior.

Players who are not inferior are generally agreed to be powerful. The conventional notion of powerless players embodied in the dummy player axiom is a quite weak one, though. In the author's view, it is too weak for a relevant class of circumstances that are modelled by simple games – in particular, if there is scope for negotiation before coalition formation and there are finitely many decisions to be taken. Under these circumstances, an inferior player i is subject to aforementioned credible ultimatum threats by some player j. The power usually associated with the swings that an inferior player may have is obliterated, and an inferior player can be expected to have only marginal influence on any economic or political decision. This suggests to strengthen the conventional dummy player axiom:

Inferior players (INF):

$(\forall v \in \mathcal{G}^I): \{i \in I^*(v) \implies \mu_i(v) = 0\}.$

As illustrated above, none of the conventional power indices satisfies the inferior player axiom.

3.3 The Strict Power Index (SPI)

In order to show that the inferior player axiom leads to reasonable power indices with desirable properties and plausible probability models, an example index related to the BZI will be developed. This section uses the traditional deterministic formulation of power indices, while Sect. 3.4 will use the probabilistic framework. Similar adaptations could be made to the SSI, the Deegan-Packel index, or other power indices. Note that bilateral ultimatum bargaining is considered in order to improve an established and frequently applied power index. This yields an adapted index with a partial non-cooperative foundation. A different line of research would use an explicit n-person bargaining model – such as the bargaining and coalition formation games investigated by Baron and Ferejohn (1989), Gul (1989), Chatterjee, Dutta, Ray, and Sengupta (1993), Hart and Mas-Colell (1996), or Okada (1996) – in order to construct new indices with a complete but highly specific non-cooperative foundation.[11] One best starts with the following adaptation of the notion of swings:

Definition 3.2. *Player i has a strict swing in winning coalition $S \in W(v)$ if*

    a) i can turn S into a losing coalition by leaving it, and
    b) i is not inferior in v, i. e. $i \notin I^*(v)$.

---

[11] Useful predictions for these n-person bargaining games are typically only possible when *stationary* SPE are considered. Osborne and Rubinstein (1990, pp. 39, 65), for example, criticize the imposition of stationary strategies, since it requires players to keep believing that some player i will make the equilibrium proposal x^* at his next move as proposer even if he has already failed to do so a thousand times.

Let

$$\tilde{\eta}_i(v) := \begin{cases} |C_i(v)|; & i \notin I^*(v), \\ 0; & i \in I^*(v) \end{cases}$$

denote the number of strict swings of player i in game v. Substituting strict swings for swings in the definition of the BZI, one gets the following new power index:

Definition 3.3. *The* strict power index (SPI) $\sigma \colon \mathcal{G}^I \to \mathbb{R}^n_+$ *is given by*

$$\sigma_i(v) := \frac{\tilde{\eta}_i(v)}{2^{n-1}}, \qquad i \in I.$$

By construction, $\sigma_i(v) = 0$ if and only if player i is inferior, and $\sigma_i(v) = 1$ if and only if i is a dictator. For the example game v_1, the SPI produces the vector $\sigma(v_1) = (\frac{3}{4}, 0, 0)$; A is the only powerful player in v_1, but still no dictator. The game v_3 with $I = \{A, B, C, D, E, F\}$ and $M(v_3) = \{ABC, ABD, ACE, BDEF\}$ illustrates that SPI and BZI index can imply different power rankings: $\sigma(v_3) = (\frac{7}{16}, \frac{5}{16}, 0, 0, \frac{3}{16}, 0)$ and $\beta(v_3) = (\frac{7}{16}, \frac{5}{16}, \frac{4}{16}, \frac{3}{16}, \frac{3}{16}, \frac{1}{16})$. C is part of smaller MWC than E. This yields a greater number of swings so that greater power is indicated by the BZI. However, C's supposed power is obliterated by his dependence on A. So, E has more strict swings that actually translate into power. Corresponding with the BZI, there is the following result:

Proposition 3.2. *The SPI satisfies (GLO), i. e. is globally monotonic.*

Proof. Consider arbitrary simple games $u, v \in \mathcal{G}^I$ with $u >_i v$. One needs to show that $\sigma_i(u) \geq \sigma_i(v)$. If i is inferior in v, this is trivial. The global monotonicity of the BZI implies $\sigma_i(u) \geq \sigma_i(v)$ if i is not inferior in u. It remains to confirm that i cannot be inferior in u without being inferior in v.

It can be verified that $u >_i v$ implies $C_i(v) \subseteq C_i(u)$. Now suppose that i is *not* inferior in v. For any player $j \neq i$, either $(\exists S_j \in C_i(v)) \colon j \notin S_j$, but then $S_j \in C_i(u)$ with $j \notin S_j$. Or $C_i(v) = C_j(v)$. Player i keeps his swings in all coalitions $S \in C_i(v)$ in game u. If either j has additional swings in u only together with i, or if there is a new coalition $S \in C_i(u)$ with $j \notin S$, the proof is finished. Otherwise, for i to become inferior in u, it must be true that a) j is part of all $S \in C_i(u)$ and that b) there is a coalition $\hat{S} \in C_j(u)$ with $i \notin \hat{S}$. $u >_i v$ implies $\hat{S} \in W(v)$. Now, one either has $\hat{S} \in C_j(v)$, which contradicts $C_i(v) = C_j(v)$. Or $\hat{S} \notin C_j(v)$, i.e. $\hat{S} \setminus \{j\} \in W(v)$. Since $\hat{S} \setminus \{j\} \cup \{i\}$ wins in v, it also wins in u. Player i cannot be crucial in $\hat{S} \setminus \{j\} \cup \{i\}$ because that would contradict a). So, $\hat{S} \setminus \{j\} \in W(u)$, contradicting b). $\qquad \square$

It can be checked that the SPI is anonymous, and hence Proposition 3.2 implies local monotonicity of the SPI. It facilitates comparisons with other power measures if an index is fully characterized by a set of logically independent axioms. An axiomatic characterization of the SPI, in the spirit of Dubey and Shapley's (1979) axiomatization of the BZI, will therefore be provided.

As mentioned above, the BZI and SSI are based on a linear notion of power which explicitly requires (ADD) to hold. For illustration, consider e. g. the set of players $I = \{A, B, C, D\}$, and games $v_4, v_5 \in \mathcal{G}^I$ with $M(v_4) = \{AB, AC\}$ and $M(v_5) = \{AD, BCD\}$. According to the BZI, B's power in $v_4 \vee v_5$ is simply the sum of power in v_4 and v_5, $\frac{1}{4} + \frac{1}{8}$, corrected by $-\frac{1}{8}$ for i's swing ABD from v_4 that becomes void due to overlap with v_5. (ADD) does not hold for the SPI: B is inferior in v_4 and v_5, but not $v_4 \vee v_5$. Therefore $\sigma_B(v_4) = \sigma_B(v_5) = 0$ is contrasted by $\sigma(v_4 \vee v_5) = (\frac{3}{4}, \frac{1}{4}, \frac{1}{4}, \frac{1}{4})$, i. e. in the composed game, B is even as powerful as player D who made B inferior in v_5. The strategic considerations underlying the inferior player axiom imply that power is additive only for special compositions and decompositions. Therefore a less restrictive requirement than (ADD) is used for the characterization of the SPI:[12]

Aggregation (AGG):
$$(\forall v \in \mathcal{G}^I): \{i \notin I^*(v) \implies \mu_i(v) = \mu_i(\bigvee_{S \in M(v)} v_S)$$
$$= \sum_{T \subseteq \wp(M(v))} (-1)^{|T|-1} \mu_i(\bigwedge_{S \in T} v_S)\}.$$

Denoting by (ABS') the straightforward adaptation of (ABS) to strict swings, the following is true:

Proposition 3.3. *The SPI is the unique power index which satisfies the four logically independent axioms (INF), (ANY), (ABS'), and (AGG), i. e. it is axiomatically characterized by them.*

Proof. (INF) and (ABS') are satisfied by construction. (ANY) follows from the anonymity of swings, and hence of strict swings. (AGG) refers to non-inferior players only. For these players, the SPI is constructed to coincide with the BZI. By complete induction, one can prove a useful lemma applying to the BZI:

Lemma 3.1. *(AGG) is satisfied by any index μ which satisfies (ADD), i. e.*

$$\mu_i(u \vee v) = \mu_i(u) + \mu_i(v) - \mu_i(u \wedge v)$$
$$\implies \mu_i(u) = \mu_i(\bigvee_{S \in M(u)} u_S) = \sum_{T \subseteq \wp(M(u))} (-1)^{|T|-1} \mu_i(\bigwedge_{S \in T} u_S).$$

Proof. Consider an arbitrary game $w^r \in \mathcal{G}^I$ with exactly $r \geq 1$ MWC, i. e. $M(w^r) = \{S_1, \ldots, S_r\}$. The claim is obviously true for $r = 1$. Proceed to $r+1$ and consider $w^{r+1} \in \mathcal{G}^I$ with $M(w^r) = \{S_1, \ldots, S_r, S_{r+1}\}$. Using additivity and the result for r, $\mu_i(w^{r+1})$ equals

$$\mu_i(w^r \vee u_{S_{r+1}}) = \sum_{T \subseteq \mathcal{P}(\{S_1, \ldots, S_r\})} (-1)^{|T|-1} \mu_i(\bigwedge_{S \in T} u_S)$$
$$+ \mu_i(u_{S_{r+1}}) - \mu_i(w^r \wedge u_{S_{r+1}}). \tag{3.2}$$

---

[12] The aggregation axiom can replace additivity in the axiomatization of BZI or SSI if its restriction to non-inferior players is dropped.

$\mu_i(w^r \wedge u_{S_{r+1}})$ is equivalent to $\mu_i\left(\bigvee_{S \in M(w^r)} (u_S \wedge u_{S_{r+1}}) \right)$. To this, the result for r can be applied once more:

$$\mu_i(w^r \wedge u_{S_{r+1}}) = \sum_{T \subseteq \mathcal{P}(\{S_1,...,S_r\})} (-1)^{|T|-1} \mu_i\left(\bigwedge_{S \in T} (u_S \wedge u_{S_{r+1}}) \right)$$

$$= - \sum_{T \subseteq \mathcal{P}(\{S_1,...,S_r\})} (-1)^{|T \cup \{S_{r+1}\}|-1} \mu_i\left(\bigwedge_{S \in T \cup \{S_{r+1}\}} u_S \right).$$

Substituting this in (3.2) proves the claim for $r+1$, and thus Lemma 3.1. □

Next, it is proved that (INF)–(AGG) uniquely define a function $\mu: \mathcal{G}^I \to \mathbb{R}_+^n$. First consider games with a single minimal winning coalition $S \subseteq I$, i. e. the auxiliary game u_S. All players $i \notin S$ are inferior in u_S and hence by (INF) $\mu_i(u_S) = 0$. All non-inferior players $j \in S$ by (ANY) have the same power $\mu_j(u_S) = a$ with $a \geq 0$. Thus, $\sum_{i=1}^n \mu_i(u_S) = a|S|$. (ABS') requires $a|S| = \frac{1}{2^{n-1}} \sum_{i=1}^n \tilde{\eta}_i(u_S)$. By construction of u_S one has

$$\tilde{\eta}_i(u_S) = \begin{cases} 0; \ i \notin S, \\ 2^{n-|S|}; \ i \in S, \end{cases}$$

implying

$$a = \frac{1}{2^{|S|-1}}.$$

Thus, μ is uniquely defined for all auxiliary games u_S with $S \subseteq I$. (INF) and (AGG) extend this definition to the entire domain $\mathcal{G}^I$.

Finally, independence of (INF)–(AGG) need to be demonstrated. The BZI β obviously violates (INF), but obeys (ABS')–(AGG). The normalized version of the SPI, $\sigma(v)/\sum_i \sigma_i(v)$, violates (ABS') but obeys the remaining axioms. An index consistent with (INF), (ABS'), and (AGG), but not (ANY) is obtained by allocating the number of strict swings in single-MWC auxiliary games to the non-inferior player with lowest order number, using (INF) and (AGG) to extend this to non-auxiliary games with multiple MWC. Indices satisfying (INF)–(ANY) that violate (AGG) will be given in Proposition 3.5 (using an appropriate re-scaling for $c \neq \frac{1}{2}$). This completes the proof of Proposition 3.3.   □

In contrast to cooperative models of spoil distribution such as the core or the nucleolus, an efficiency requirement makes only little sense when power is concerned. A normalization such that $\sigma(\cdot)$ adds up to 1 is therefore not necessary. It would, moreover, destroy its global monotonicity,[13] its probabilistic foundation (see next section) and, in fact, an important part of the information that is given by the SPI.

---

[13] Consider $I = \{A, B, C, D, E, F\}$ and $v_6, v_7 \in \mathcal{G}^I$ with $M(v_6) = \{ABC, ABDE, ACDF, BCDEF\}$ and $M(v_7) = \{ABC, ABDE, ACDF, BCEF\}$. Normalizing $\sigma(\cdot)$ yields $(\frac{11}{34}, \frac{9}{34}, \frac{9}{34}, \frac{5}{34}, 0, 0)$ for v_6 and $(\frac{1}{3}, \frac{1}{3}, \frac{1}{3}, 0, 0, 0)$ for v_7 – indicating that A has more power in v_7 although $v_6 >_A v_7$. The author thanks René Levínský for suggesting this example.

3.4 Inferior Players in a Probabilistic Setting

In the probabilistic setting, the property of player $i \in I^*(v)$ being an inferior player has to be reflected in some way by i's acceptance rate p_i. One can find a plausible restriction on p_i by recalling that inferior players have to content themselves with essentially a zero share of economic or political spoils when belonging to a winning coalition. This means that an inferior player is basically indifferent between joining a winning coalition or staying outside, between voting for or against a proposal. This can be formalized by:

Strict Power Condition (SPC): i is inferior in $v \implies p_i \equiv \frac{1}{2}$.

One gets the following probabilistic foundation of the SPI:

Proposition 3.4. *Applying the SPC in the setting of the probabilistic BZI, i. e.*

$$p_i \begin{cases} \equiv \frac{1}{2}; & i \in I^*(v), \\ \overset{i.i.d.}{\sim} U[0,1]; & i \notin I^*(v), \end{cases}$$

implies the probabilistic SPI.

The proposition follows from the more general Proposition 3.5 below ($c = \frac{1}{2}$). Note that imposition of the SPC changes the interpretation of power polynomial $f_i(p_1, \dots, p_n)$. It no longer gives the probability of player i having a swing or being crucial in the random coalition that is to be formed, but the probability of player i having a strict swing or of being crucial in a way that actually permits exertion of power.

Figure 3.1 illustrates example game v_1 and the SPC. The deterministic winning coalitions are indicated by black corner points and losing coalitions by white corner points of the unit cube. All other points of the cube correspond to fuzzy or random coalitions. Calculation of the BZI involves taking expectations of the derivative of v_1's multilinear extension over the entire cube. The SPC restricts the domain of v_1's MLE to the broken line.

Inferior players' practical indifference towards being part of a winning coalition can, of course, be formalized differently. For example, one could assume that inferior players join whatever coalition is decided on by the powerful players of the game with probability one, or probability zero, or some probability c in between. This leads to the

Generalized Strict Power Condition (GSPC): i is inferior in $v \implies p_i \equiv c, \ c \in [0,1]$.

The GSPC restricts the domain of v's MLE to the $(n-m)$-dimensional unit cube, where $m = |I^*(v)|$ denotes the number of inferior players in v. In order to characterize those deterministic indices whose probabilistic counterpart satisfies the GSPC for some $c \in [0,1]$ one needs to decompose $\tilde{\eta}_i(v)$ and generalize the notion of strict swings.

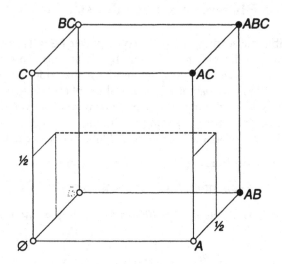

Fig. 3.1. Illustration of v_1 and the strict power condition (SPC)

Definition 3.4. *Player i has a θ-swing in winning coalition $S \subseteq W(v)$ if*
a) i can turn S into a losing coalition by leaving it,
b) i is not inferior in v, i. e. $i \notin I^*(v)$, and
c) exactly θ inferior players are part of S.

Let

$$\eta_i^{(\theta)}(v) := |\{S \subseteq I : S \in C_i(v) \,\wedge\, i \notin I^*(v) \,\wedge\, |S \cap I^*(v)| = \theta\}|$$

denote the number of θ-swings of player i in game v. One trivially has

$$\sum_{\theta=0}^{m} \eta_i^{(\theta)}(v) = \tilde{\eta}_i(v).$$

Various indices can be defined based on the primitive θ-swing. Averaging $\eta_i^{(\theta)}(v)$ with particular weights on each θ can incorporate especially plausible or empirically relevant assumptions about inferior players' behaviour. A continuum of anonymous power indices which satisfy the inferior player axiom can now be probabilistically characterized:[14]

Proposition 3.5. *A MLE satisfying the GSPC gives zero power for inferior players. Applying the GSPC in the setting of the probabilistic BZI, i. e.*

$$p_i \begin{cases} \equiv c; & i \in I^*(v), \\ \overset{i.i.d.}{\sim} U[0,1]; & i \notin I^*(v) \end{cases}$$

---

[14] Absolute power axiom (ABS') is satisfied by appropriate re-scalings.

for some $c \in [0, 1]$, implies the generalized strict power index (GSPI) σ^c *with*

$$\sigma_i^c(v) := \sum_{\theta=0}^{m} c^{\theta}(1-c)^{m-\theta} \frac{\eta_i^{(\theta)}(v)}{2^{n-m-1}}$$

Proof. W. l. o. g. consider $v \in \mathcal{G}^I$ with non-inferior players $1, \ldots, n - m$. If v's MLE $f(p_1, \ldots, p_n)$ satisfies the GSPC, then it is a non-degenerate function only of $p_1, \ldots, p_{n-m}$. Thus $\partial f(p_1, \ldots, p_{n-m})/\partial p_i = 0$ for $i > n - m$.

Imposing the GSPC to a MLE gives

$$f(p_1, \ldots, p_n) = \sum_{\substack{S \in W(v)}} \prod_{\substack{i \in I(v) \\ i \in S}} c \prod_{\substack{j \in I(v) \\ j \notin S}} (1-c) \prod_{k \in S \setminus I^*(v)} p_k \prod_{l \notin S \cup I^*(v)} (1-p_l)$$

$$= \sum_{S \in W(v)} c^{\theta(S)}(1-c)^{m-\theta(S)} \prod_{k \in S \setminus I^*(v)} p_k \prod_{l \notin S \cup I^*(v)} (1-p_l),$$

where $\theta(S)$ indicates the number of inferior players in coalition S. Taking expectations of the partial derivative with respect to p_i for $i \notin I^*(v)$ yields

$$Ef_i(p_1, \ldots, p_n) = \sum_{S \in C_i(v)} c^{\theta(S)}(1-c)^{m-\theta(S)} \left(\frac{1}{2}\right)^{n-m-1}.$$

Considering only one $S \in C_i(v)$ for each number $\theta \in \{0, \ldots, m\}$ of inferior members and weighting this summand by the number $\eta_i^{(\theta)}(v)$ of such coalitions in $C_i(v)$ produces the claim. $\qquad\square$

The SPI is the special case of $c = \frac{1}{2}$. The special case of $c = 1$ corresponds to the *Follower-Leader Index of Power (FLIP)* defined in Napel and Widgrén (2000). There, the requirement $p_i \equiv 1$ for $i \in I^*(v)$ is slightly weakened to the restriction that $p_i p_j \equiv p_j$ if $C_i(v) \subsetneq C_j(v)$. This asks for a stronger type of behavioural similarity than the correlation assumption of the SSI: The perfect correlation of acceptance rates means that inferior players follow the common standard t but take decisions independently of the non-inferior players. In contrast, $p_i p_j \equiv p_j$ means that the inferior player i unequivocally supports proposals which have a positive chance of being supported by player j. Inferior players thus follow the non-inferior 'leaders' of the game into whatever coalition they decide on. The FLIP is suited to environments in which the inferior players are especially benign.

3.5 Concluding Remarks

Motivated by the discrepancy between power indications given by, on the one hand, established indices that are based on the dummy player axiom and, on the other hand, the important aspect of power captured by ultimatum bargaining, this chapter has argued in favour of strengthening the commonly used dummy player axiom of power measurement. The inferior player axiom has been proposed as a substitute which is based on bilateral bargaining theory.

In order to demonstrate that meaningful indices which comply with the inferior player axiom can be constructed, the strict power index (SPI) has been proposed. It has first been analysed in a traditional deterministic setting and axiomatized. For a comprehensive understanding of the concept of inferior players, its probabilistic counterpart then has been investigated; a probabilistic condition that implies the SPI has been derived, and generalized.

Future research may apply the inferior player axiom to other indices than the non-normalized Banzhaf index, e.g. those of Shapley and Shubik or of Deegan and Packel. It could be worthwhile to investigate more thoroughly the mathematical properties of the respective adaptations of the Banzhaf, Shapley-Shubik or Deegan-Packel indices in terms of axiomatization, monotonicity, and susceptibility to so-called paradoxes in power measurement (cf. Felsenthal and Machover 1998, ch. 7, for a discussion of the latter). The inferior player axiom could be extended to the domain of general games in characteristic function form. One can also define a related stability concept for coalition structures, i.e. partitions of the set of players. $I(nferior\ coalition)$-$stability$ would require of a partition $\{S_1, \ldots, S_q\}$ of I that no element is inferior in the reduced game among coalitions $S_1, \ldots, S_q$. For example, the I-stable structures in v_1 are $\{A, BC\}$ and $\{ABC\}$. The relationship to the stability notions behind the bargaining set of Aumann and Maschler (1964) and similar concepts is yet unexplored. Moreover, one could explicitly take into account who makes a given player i inferior in order to define a partial ordering of players rather than only a partition of I.

A common criticism concerning the power index approach which has been focussed on in this chapter stems from the fact that it ignores players' preferences and institutional factors like agenda setting (e.g. Garret and Tsebelis 1999). The obvious counter-argument to this criticism is that power indices are measures of $a\ priori$ power. This kind of analysis is needed, for example, to design voting rules in a new constitution (see Holler and Widgrén 1999) or to guide institutional amendments e.g. in the process of enlarging the European Union (Baldwin, Berglöf, Giavazzi, and Widgrén 2000). It is of limited use when a parliament with given party preferences and seat distribution is to be analysed.

By their expectation character, clearly visible in Sect. 3.4, power indices do typically not measure actual short-run realizations of power.[15] This is in

---

[15] Power indices have, however, been successfully applied as a proxy for long-run average realizations of power. For example, Baldwin, Francois, and Portes (1997) investigate the hypothesis that the EU budget is distributed in line with the

particular the case when players have preferences which imply that actual decisions rather than only acceptance rates are dependent across players. Then the multilinear extension becomes an inappropriate tool of analysis, and so do the indices founded on it; the above framework of simple games is no longer sufficient.

Recent work on power measurement takes spatial preferences and institutional aspects of power like agenda-setting or other specific decision procedures more explicitly into account, but tries to retain some crucial aspects of a priori measures of power. Examples are e. g. Steunenberg, Schmidtchen, and Koboldt (1999) or Widgrén and Napel (2001). In the former, power is measured as the average distance between players' ideal points and equilibrium outcomes in the considered policy games. This ignores the, in the author's opinion, important distinction between the mere *luck* of having preferences matching an outcome and the *power* to bring about a desired outcome. In Widgrén and Napel (2001), only the latter is considered relevant. *Spatial inferiority* is defined in a 1-dimensional spatial voting model with an agenda setter. A player is called spatially inferior if his random ideal point almost never affects the subgame perfect policy proposal of the agenda setter. Spatial inferiority implies inferiority, but the reverse is not true.

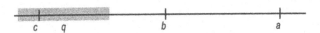

Fig. 3.2. Example v_1 and spatial preferences

For illustration, let a, b, and c denote A's, B's, and C's respective deterministic, commonly known ideal points for above example of a federal and two regional governments, v_1. Let q denote the status quo, which is imagined to lie in the grey region indicated in Fig. 3.2. If regional government C's ideal point is to the left of q, then C prefers the status quo to any policy proposal that A will rationally make. Since there are no spoils associated with coalition AC or gains from 'trade', the statement that C has no or only marginal influence on the 'terms of trade', which is appropriate in a simple game setting, loses its meaning. Moreover, spatial preferences imply that A is generically no longer indifferent whether to make an offer to B or C. This will affect the extent to which players B and C have to be content with marginal vs. substantial concessions from A. Taking A to be the natural agenda setter in this spatial voting game, the expected impact and thus power of B depends on the distribution of all players' ideal points and the status quo. In the preference constellation of Fig. 3.2, it depends on q whether B's ideal point b has

---

distribution of power in the Council of Ministers. They, in fact, find a high correlation between SSI per capita and EU receipts per capita among member states.

an effect ($x^* = 2b - q$ if $b - q < a - b$) or not ($x^* = a$ if $b - q \geq a - b$) on A's optimal policy proposal. Considering a random status quo and random ideal points establishes a novel link between spatial voting models and traditional power analysis via simple games.

4. Bargaining and Justice

Bargaining situations are usually analysed in terms of an ordinal or cardinal representation of players' preferences, and not the underlying allocations of goods, monetary transfers, etc. This abstraction makes bargaining models versatile and applicable in many contexts. In Chap. 1, the classical bargaining problems of bilateral exchange and the division of rents in a bilateral monopoly have been the chief examples. Chapter 2 has considered the repeated division of surpluses which may also be created by social rather than economic interaction. Then, Chap. 3 has investigated implications of bilateral bargaining in the context of political decision-making. This chapter will depart further from pure economics. It deals with principles of social justice, and the norms and institutions which organize surplus sharing at the level of society.

John Rawls observes in his *Theory of Justice* (1971, p. 4) that

> ... although society is a cooperative venture for mutual advantage, it is typically marked by a conflict as well as by an identity of interests. There is an identity of interests since social cooperation makes possible a better life for all ... There is a conflict of interests since persons are not indifferent as to how the greater benefits produced by their collaboration are distributed ...

Unless a society is to resolve every single instance of the bargaining situations described by Rawls via explicit face-to-face negotiation, constitutional arrangements, political rules, or social conventions which determine divisions of advantages are required. Various *distribution norms* are candidates to serve as the construction principle of such institutions.[1] They can abstractly be represented as cooperative bargaining solutions.

The objective of this chapter is, first, to analyse the distinct principles of justice that would be reflected by implementing the bargaining solutions introduced in Sect. 1.2.[2] This, in Sect. 4.1, identifies the normative, usually implicit arguments behind different bargaining solutions. For example, the common misconception of the Nash bargaining solution being designed as

---

[1] Real societies' legal and political institutions seem to be rarely derived from coherent first principles. Here, an 'ideal' society is considered which tries to deduce more specific concepts of social justice and arrangements such as property rights, competitive markets, public education, or military service from a single general norm of social surplus distribution.

a fair arbitration scheme is clarified. Then, arguments for the adoption of a particular conception of justice, understood mainly as a particular bargaining solution, are investigated and evaluated. Section 4.2 deals with the seminal analysis of Rawls (1971). Section 4.3 discusses the recent contribution by Binmore (1994, 1998b). Section 4.4 evaluates the advantages and limits of approaching questions of social philosophy with game-theoretic models. The presentation draws on Napel (1998, 1999a) and Holler and Napel (2001).

4.1 Bargaining Solutions and Principles of Social Justice

Economists typically try to avoid *interpersonal comparison of utility*. A main reason is that such comparisons are difficult to justify within the positivistic paradigm of revealed preference. Also, there exists the view that economics should leave the discrimination between efficient distributions to philosophy or political science. However, there is little controversy that, first, people do compare the well-being of different persons in many situations – e. g. in the form of envy and anger when they feel unjustly disadvantaged themselves, but also sympathy when others are unfairly treated – and, second, that this affects their decision making. Comparison of subjective well-being seems to be a part of human nature that becomes relevant as soon as individuals get connected within a social space. Harsanyi (1975, p. 600) holds that inter-personal comparison is "a concept we cannot avoid in any reasonable theory of morality."[3] It seems that a discussion of justice and fairness cannot do entirely without the comparability assumption. A possible formalization of it is given along the lines of Harsanyi (1977, ch. 4) in Sect. 4.3 (drawing on the presentation by Binmore 1994, pp. 282ff). Two types of comparability will be considered, namely the possibilities of the comparison of *differences* or *increments* of player i's utility with those of player $-i$, and of the comparison of the actual utility *levels* of the players (see e. g. Harsanyi 1987 for details on this distinction).

4.1.1 Nash Bargaining Solution

Nash's bargaining solution has been introduced in Chap. 1.2. It is highly prominent in economic applications. The popularity of the Nash solution, on

---

[2] All mentioned cooperative solutions can easily be generalized to a society com-prising n rather than two individuals as long as unanimous agreement is required and coalition formation has no bearing.

[3] Harsanyi defends interpersonal utility comparisons by arguing that "preferences and utility functions of all human individuals are governed by the same basic psychological laws" (p. 600). With this assumption, comparison of utility reduces to the evaluation of distinct sets of social, physical, and economic positions by a universal meta-level utility function.

the one hand, stems from its particularly sound theoretical foundation as an approximation of the outcome predicted by various bargaining models based on both highly rational, perfectly informed agents (cf. Rubinstein's alternating offers model in Sect. 1.3.2) and also very boundedly rational, imperfectly informed agents (see e. g. Young's model of adaptive play in Sect. 1.4.1). On the other hand, the Nash solution F^N and its asymmetric version $F^{N(a,b)}$ have a particularly simple and manageable functional form. From its first proposal by Nash (1950a), F^N has often been considered a fair arbitration scheme, associated with the selection of a fair bargain. Luce and Raiffa (1957, pp. 128–129) write

> Nash contends that his solution is a 'fair' division which purportedly should reflect the 'reasonable expectancies' of 'rational bargainers.' ... assumptions 1 and 4 [axioms (INV) and (SYM)] stipulate the principle of 'fairness.'

They clarify that "Nash's solution only purports to give a 'fair' arbitrated value of the bargaining when the *strategic aspects* are taken into account" (p. 130, italics in the original) but maintain the supposed link between F^N and fairness. Bishop (1963, p. 563) similarly claims to discuss " 'axioms' which Nash regards as 'fair' and 'reasonable' conditions." Many direct or indirect associations of Nash's solution with fairness can be found elsewhere, e. g. in Rosenmüller (1981, p. 373) or Holler and Illing (2000, p. 191). Nash himself is not very explicit on the topic. He states (Nash 1950a, p. 158, italics added)

> ... we may think of one point in the set of the graph as representing the solution, and also representing all anticipations *that the two [players] might agree upon as fair bargains* ... the 'fair bargain' might consist of an agreement to use a probability method ...

and later (Nash 1953, p. 136 – note the absence of a reference to fairness)

> ... one can attack the problem axiomatically by stating general properties that 'any reasonable solution' should possess.

These statements do not contradict the above conjectures about Nash's contentions, but are not overly supportive either. In any case, as will be seen below, there is only a *pragmatic* basis for the use of the Nash solution as a society's rule for *fair* or *just* surplus division.

F^N's explicit invariance to equivalent individual utility representations, (INV), formalizes that Nash does not assume any interpersonal comparability of utility. To the contrary, (INV) implies that comparisons between prescribed utility levels $F_i^N(U, u^D)$ and $F_{-i}^N(U, u^D)$ are considered as meaningless.[4] Note well that this is equally true if the bargaining problem $\langle U, u^D \rangle \in \mathcal{B}$ happens to be symmetric. Nash's symmetry axiom, (SYM), is a natural mathematical requirement; it does *not* formalize any type of fairness or egalitarian principle of justice.

---

[4] Basically the same arguments apply to the asymmetric Nash solution $F^{N(a,b)}$. For brevity, only the Nash solution F^N will be discussed here.

The Nash solution neglects all details of good exchange, wage setting, etc. underlying a bargaining problem $\langle U, u^D \rangle \in \mathcal{B}$. It therefore cannot be expected to implement first principles of fairness which refer to these details. A prominent such concept is the *envy-freeness criterion*. It is frequently applied when interpersonal comparisons of utility are not (regarded as) possible.[5] An allocation of goods, money, etc., represented by a vector x and associated with a given bargaining outcome $o \in O$, is said to be *envy free* if each player personally does not prefer the bundle of goods, money, etc. of another player to his own, i.e. $(\forall i \in I)$: $\pi_i(x_i) \geq \pi_i(x_{-i})$. So, all comparison is with respect to objective quantities. F^N does not satisfy envy-freeness: As seen in Sect. 1.2.1, a risk-averse player receives less than 50 cents of a euro which is to be divided with a risk-neutral player – making the former envious of the latter's share.

Though the Nash solution does not assume interpersonal comparability of utility, one may still suppose that comparisons of levels or increments are possible for given players. Consider e.g. two heterogeneous firms operating in a regulated market; some aspects of regulation, such as work safety or environmental standards, affect the distribution of rents from imperfect competition. Assume that a units of firm 1's share of rents are socially comparable to b units of firm 2's share, and let the set of feasible payoff combinations – corresponding to particular social contracts – offer intrinsically better opportunities to firm 1 than to firm 2. A procedure based on the principle of equity would treat both firms in some sense equally based on ratio a/b.[6] The Nash solution is not designed to satisfy such a rule. To see that it does in fact not, note that by the independence of irrelevant alternatives, (IIA), and (INV), it treats any bargaining problem $\langle U, u^D \rangle$ exactly as a symmetric problem $\langle U', 0 \rangle$: First, (IIA) requires F^N to assign the same outcome $o \in O$ to $\langle U, u^D \rangle$ as it does to the problem $\langle \tilde{U}, u^D \rangle$ where $\tilde{U}$ has a linear Pareto-boundary and adds only irrelevant alternatives to U. Second, affine transformations τ_i can be applied to $\tilde{U}$ in order to produce the symmetric problem $\langle U', 0 \rangle$. (SYM) and Pareto efficiency, (PAR), then determine the by (INV) identical outcome of the transformed problem $\langle U', 0 \rangle$ and of the original problem $\langle U, u^D \rangle$. (SYM) imposes an equal treatment of both players in utility space – even an equal final utility for both players – but in the terms a'/b' of utility comparison resulting from the symmetrization. Since the symmetrizing transformations τ_1 and τ_2 are usually different, ratios a'/b' and a/b generally differ. So, (SYM) – in conjunction with (IIA) and (INV) – treats both players equally in terms of a quite arbitrarily distorted standard.

As seen in Sect. 1.2.1, the Nash solution – independently of the chosen von Neumann-Morgenstern utility representations – equalizes something, namely players' respective *risk limit*. It denotes the maximum probability of

---

[5] Compare Varian (1974) and the fair division procedures studied by Brams and Taylor (1996).

[6] This does not imply an egalitarian outcome.

conflict that a player is prepared to face when insisting on better than the offered terms (cf. Zeuthen's behavioural model of alternating concessions, p. 9). Hence the one and only type of equality which can be associated with F^N is an equality in players' potential to fight for better terms. The Nash solution implements a balance of power which is corroborated by Zeuthen's, Rubinstein's, and Young's bargaining models.

Thus, a society whose norm for distributing gains from social cooperation corresponds to the Nash bargaining solution can be said to equate justice with power, i. e. it is deemed just, what is stable given the society's distribution of power.[7] The Nash solution's efficiency promotes overall efficiency in society. Moreover, it is advantageous that the stability of an F^N-based norm is founded on the realities of power, and not only the fact that any efficient social contract is better than none: People will find it in their interest to challenge – at possibly considerable social transaction costs – society's institutions and arrangements only if the balance of power has shifted to their favour.

4.1.2 Kalai-Smorodinsky Bargaining Solution

The Kalai-Smorodinsky solution has been introduced in Sect. 1.2.2. It is characterized by axioms (PAR), (SYM), (INV) and individual monotonicity, (IM). The statements about F^N which are based on (INV) and (SYM) therefore apply to F^{KS}, too. In particular, the Kalai-Smorodinsky solution does not select envy-free outcomes. For example, a risk-averse player 1 with von Neumann-Morgenstern utility function $\pi_1(x_1) = \sqrt{x_1}$ receives about 0.38 cent when a euro is to be split with a risk-neutral player 2, i. e. $\pi_2(x_2) = x_2$, and initial endowments are zero – an obvious cause of envy.

However, by definition, F^{KS} assigns to each player exactly the same percentage achievement of his maximal possible gain on the status quo, i. e. $u^* = F^{KS}(U, u^D)$ satisfies[8]

$$\frac{u_1^* - u_1^D}{u_1^B(U) - u_1^D} = \frac{u_2^* - u_2^D}{u_2^B(U) - u_2^D}. \tag{4.1}$$

Since in above example both players have the same ideal utility level, $u_i^B(U) = 1$, and the status quo is $(0,0)$, the asymmetric Kalai-Smorodinsky distribution $x^* \approx (0.38, 0.62)$ corresponds to equal percentage achievements of maximal gains and also to equal final utility $\pi_1(x_1^*) = \pi_2(x_2^*)$. The Nash solution's division is approximately $(0.33, 0.67)$, having no such property.

Equation (4.1) points out an egalitarian flavour of the Kalai-Smorodinsky solution. Still, F^{KS} implements not equality of outcomes but of realized potential:[9] If a bargaining problem arising in society is such that player i has

---

[7] There is also a slightly different interpretation of the equalization of players' risk limits: It is deemed just, what corresponds to players' willingness to risk conflict.

[8] The denominators are necessarily positive by the assumptions on $\langle U, u^D \rangle \in \mathcal{B}$.

greater opportunities or would have greater maximum gains from cooperation than everybody else, he will be assigned the most (assuming that a standard of comparison is defined). F^{KS} as a norm takes distributional justice to mean equal success of everybody's striving for maximal personal happiness in social interaction. It is deemed just, what is equally imperfect from everybody's point of view, or, what realizes everybody's paradise to the same degree. Here, 'maximal personal happiness' or 'paradise' are defined relative to status quo u^D. What is the appropriate choice for it, e. g. in the context of reforming institutions of a not-so-ideal society, is a difficult question. Options in line with the discussion concerning the correct choice of u^D in the context of the Nash solution in Sect. 1.3.2 (p. 43) include cooperation according to the present social contract as well as a 'natural state' with no social cooperation and, possibly, even Hobbesian war of all against all. Recall that the straightforward extension of F^{KS} to $n \geq 3$ players violates (PAR). Possible modifications of F^{KS} which guarantee efficiency cannot guarantee players' equal percentage realization of maximal possible gains.

4.1.3 Egalitarian and Utilitarian Bargaining Solutions

The egalitarian and the utilitarian bargaining solutions, F^E and F^U, have been introduced in Sect. 1.2.2, along with their asymmetric or weighted variants. They implement two distinct first principles of social justice which require that coherent interpersonal comparison of utility levels or utility increments is possible.

When one incremental unit of player i's utility corresponds to one incremental unit of player $-i$'s utility, the egalitarian solution F^E distributes all gains, which are available from social cooperation relative to the status quo u^D, equally among the members of society, I. If u^D describes an initial state of equality or, specifically, of equal comparable utility levels, then $F^E(U, u^D)$ assigns the same final utility to every player $i \in I$. This rule leaves possible gains unexploited if they would imply inequality, i. e. the egalitarian solution does not satisfy Pareto efficiency (see Fig. 4.1[10]).

The weighted egalitarian solution, also known as the proportional solution, $F^{E(a,b)}$, has two different distributional meanings. In the first case, $F^{E(a,b)}$ readjusts player 1's utility units by the factor $1/a$ and player 2's utility units by $1/b$ because the players' utility scales differ from the comparable standard – expressed in 'social utils' – by factors a and b, respectively. Then,

---

[9] The alternative definition of the ideal point given on p. 22, fn. 19, can be interpreted as ensuring that only that potential is considered relevant which does not conflict with other players' basic rights formalized by the status quo.

[10] Note that U is not u^D-comprehensive, which would be ensured by players' ability to freely dispose of 'excess utility.' In this case, $F^E(U, u^D)$ is at least weakly efficient. Free disposability cannot be taken for granted, however, especially if utility represents general revealed preferences and not simply a monotonic transformation of money, for example.

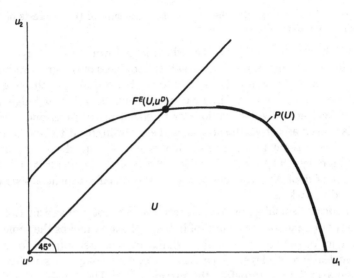

Fig. 4.1. Inefficiency of the egalitarian solution F^E

it acts like F^E. In this case, F^E and $F^{E(a,b)}$ implement that it is deemed just, what gives equal happiness to everybody.

A second possibility is that individual utility functions π_i are already expressed in social utils. The application of factors $1/a$ and $1/b$ then implies that some individuals are as in Orwell's *Animal Farm* considered 'more equal' than others. The scaling factors can be said to reflect the distinct *worthiness* of different members of society; a great a formalizes great worthiness of player 1. The weighted egalitarian solution $F^{E(a,b)}$ then formalizes that it is deemed just, what is proportional. This principle is traced back by Binmore (1998b, p. 397) to Aristotle's *Nicemachean Ethics*; a number of empirical references to its support are listed in Binmore (1998a). A society's use of $F^{E(a,b)}$ means that surplus-distributing institutions are constructed in a coherent way. However, without a specification of the worthiness factors or how they are to be determined, the proportional principle of social justice is vacuous. With appropriate weights, democracy is equally in line with $F^{E(a,b)}$ as are absolutistic monarchy or dictatorship.

The utilitarian bargaining solution F^U implements the distribution of social gains espoused by *utilitarianism*. One of its first proponents, Jeremy Bentham, summarizes its central moral and political tenets by describing the "principle of utility" – later also named the "greatest happiness principle" – as (Bentham 1789, ch. I, §§ 2–4)

> ... that principle which approves or disapproves of every action whatsoever, according to the tendency which it appears to have to augment or diminish the happiness of the party whose interest is in question ... The

interest of the community then is, what? – the sum of the interests of the several members who compose it.

More specifically, it is "a principle, which lays down, as the only *right* and justifiable end of Government, the greatest happiness of the greatest number" (addendum made by Bentham in 1822 to ch. I, § 8, italics in the original).[11] The greatest happiness principle can yield conflicting moral judgements if applied to either individual actions or to social institutions and rules. The former is known as *act utilitarianism*, and e. g. evaluates it to be just if a poor person does not repay his debt to a rich person (assuming decreasing marginal utility of money). The latter is known as *rule utilitarianism*; it calls for the repayment of debt as a rule necessary for the socially beneficial existence of liquid credit markets.

F^U allocates social gains to maximize the sum total of satisfaction on the premise that one incremental unit of individual utility means the same among all members of society. $F^{U(b,a)}$ does similarly, but first adjusts for different utility scales across players – equating a units of player 1's utility with b units of player 2's utility before the summation.[12] The status quo u^D affects the utilitarian solution only as a constraint on feasible utility combinations. Maximizing the sum or the average of a finite and fixed number of quantities is equivalent. Therefore, assuming a constant population I, F^U and $F^{U(b,a)}$ are the abstract representation of the social principle which deems just, what achieves maximal average happiness.

This takes distributional aspects only indirectly into account. When individuals experience decreasing marginal utility from their surplus share, a tendency towards the equal distribution exists.[13] In contrast to the principle of justice formalized by the egalitarian solution, it is acceptable according to F^U to compensate one person's loss with another person's gain. For example, it would be considered just if the poor get poorer provided that the rich get sufficiently richer. Rawls (1971, pp. 26ff) observes – and criticizes – that this translates to society's choice the tenets of rational individual choice, allowing negative payoff in some period (for some person) in order to maximize total utility over time (in society). Although Rawls notes the appeal of "maximizing something" (p. 25), he claims that by this feature "utilitarianism does not take seriously the distinction between persons."

---

[11] This is written in the context of rapid political changes occurring in Bentham's time (1748-1832) and, in particular, the contrasting type of government "which has for its *actual* end or object the greatest happiness of a certain *one*" (addendum to ch. I, § 8, italics in the original).

[12] If players' utility is already measured in the same units, a and b reflect interpersonal discrimination factors – e. g. interpreted as worthiness – exactly as for weighted egalitarian solution.

[13] If everybody has the *same* strictly concave utility function, an egalitarian outcome results. Bentham has explicitly and repeatedly argued that individual pleasure exhibits diminishing returns (cf. Spiegel 1991, p. 343). Moreover, his summation instructions (ch. IV, § 5 (6)) specify that every person is to count for one, and no one for more than one. This calls for a rather egalitarian distribution.

4.2 Rawls's Theory of Justice

Rawls's (1971, p. 10) professed aim is to consider the implications of "a conception of justice that nullifies the accidents of natural endowments and the contingencies of social circumstance" (p. 15) – with the hope of being led to a different conclusion than utilitarianism.[14] The main analytical construct for his investigation is a hypothetical *original position* in which members of society are imagined to meet in order to rationally and freely negotiate, once and for all, a binding social contract which specifies what counts as just and unjust among them.

The crucial feature of bargaining in the original position is that no one knows his place in society – neither his class position, wealth, and other social circumstances, nor his intellectual talents, physical abilities, and possible disabilities. Even actual personal preferences are clouded by a metaphorical *veil of ignorance*.[15] Each player bargains using meta-preferences which capture his personal interests in future life, taking into account that he might end up in any realization of social positions, preferences, etc. after the final deal is struck. Rawls calls the fundamental agreements reached under these circumstances *fair*. Since he tries to identify the principles of justice which would be specified by such a fair bargain, he labels his theory *justice as fairness*.

Once the principles of justice are determined in Rawls's model society, every individual is supposed to support the institutions and social arrangements thus defined as just. If someone ends up in an undesirable social position, renegotiation is not feasible. This fact is anticipated in the original position. It implies that players are not restricted to make agreements which will be self-policing in terms of personal preferences after the veil of ignorance has been lifted.

One way to deduce which distribution norm will be agreed on in above setting is to apply Bayesian decision theory and bargaining theory.[16] In particular, each player's meta-preferences in the original position could be assumed to have a von Neumann-Morgenstern form. All personal preferences then need to be weighted by the respective probability of having them, and applied to the set of feasible social contracts. If players try to maximize this expected utility in the original position, they are in fact maximizing a convex combination of the utility functions of all members of society. Thus, every player precisely uses meta-preferences corresponding to a weighted utilitarian solution; equiprobable social positions correspond to the utilitarian solution.

[14] Rawls has later withdrawn from several of the positions espoused in Rawls (1971). A recent update is given in Rawls and Kelly (2001). This section focusses on his theory of justice as it was originally laid out.

[15] Rawls (1971), in fact, considers several different veils – or, equivalently, a veil whose thickness is diminished in several steps (p. 200).

[16] This is probably the natural way for an economist – though not a naturalistic way. The argument of this paragraph will be stated more formally in Sect. 4.3.

If everyone has the same meta-preferences, the utilitarian social norm will be the fair bargain.

Rawls, however, does not use Bayesian decision theory in his analysis. He concludes that the following two principles will be agreed on by the bargainers (Rawls 1971, p. 60):

> First: each person is to have an equal right to the most extensive basic liberty compatible with a similar liberty for others.
> Second: social and economic inequalities are to be arranged so that they are both (a) reasonably expected to be to everyone's advantage, and (b) attached to positions and offices open to all.

The two principles have a lexicographic priority, i. e. no amount of everyone's economic advantage can compensate an infringement of someone's basic liberty rights. Part (a) of the second principle is clarified to mean that the so-called *difference principle* should be applied, stating that (p. 75):

> ... higher expectations of those better situated are just if and only if they work as part of a scheme which improves the expectations of the least advantaged members of society.

Improving the expectations of the most disadvantaged can be understood in a strict or weak sense. The latter ensures that the difference principle is compatible with efficiency; the point R in Fig. 4.2 would accordingly be selected by the fair bargain reached in the original position.[17] The corresponding function which maps general bargaining problems $\langle U, u^D \rangle \in \mathcal{B}$ to utility combinations $u^* \in U$ will be called the *Rawlsian (bargaining) solution* and denoted by F^R. Formally, $F^R : \mathcal{B} \to U$ can be defined by

$$F^R(U, u^D) := \max \left\{ \operatorname*{arg\,max}_{\substack{u \in U \\ u \geq u^D}} \left\{ \min_{i \in I} u_i \right\} \right\},$$

where the first maximization is with respect to the $\geq$-partial ordering of real vectors.[18] The status quo u^D enters only as a constraint. The objective function considers utility levels, so F^R requires comparability of levels, not only of increments. A weighted variant of F^R, which first re-scales players' utilities, can analogously be defined.

The difference principle, implemented by F^R, is usually referred to as the *maximin equity criterion* (short for 'maximum minimorum'). Rawls (1974, p. 141) stresses that it and the maximin rule of choice under uncertainty are "very different things." But in Rawls's setting the former's formal derivation

---

[17] The rectangular social indifference curves shown in Fig. 4.2 correspond to the strict improvement interpretation. The weak interpretation allows that the tangency point with greatest coordinates be selected.

[18] This ensures efficiency and uniqueness in the 2-player case. In an n-player setting, one would maximize utility of the second-worst-off player subject to the worst-off player's utility being maximal, then maximize utility of the third-worst-off player, etc. This is also known as the *leximin equity criterion*.

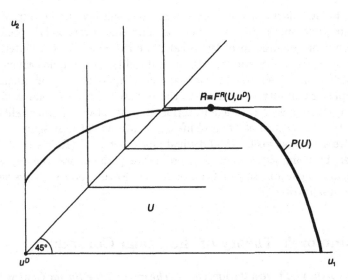

Fig. 4.2. Rawlsian solution of bargaining problem $\langle U, u^D \rangle$

as the result of bargaining in the original position rests on players' use of the latter. Free and rational players want to maximize the utility of the least-advantaged person only if everyone acts as if this were the only position which matters to further *his* personal future interests, e. g. as if each bargainer was sure to actually receive that position.

Rawls (1971, p. 153) acknowledges that "the maximin rule is not, in general, a suitable guide for choices under uncertainty"[19] but defends its application "in situations marked by certain special features." He holds these features to be an impossibility to know the probability distribution over the possible states of nature, satisficing preferences which exhibit non-positive marginal utility once a certain aspiration level is attained, and alternatives which have "outcomes that one can hardly accept" (p. 154), corresponding to infinitely negative utility e. g. if one is a slave or serf. According to Rawls the "original position clearly exhibits these special features" (p. 157).

Arguing against utilitarian conclusions that can be drawn from Rawls's very setting as indicated above – espoused most notably by Harsanyi (1975, 1977)[20] – Rawls (1974) later concentrates on defending not so much players' use of the maximin choice rule in the original position but rather the resulting maximin equity criterion. The arguments he gives in support of this principle

---

[19] The maximin rule is, for example, incompatible with the maximization of a von Neumann-Morgenstern utility function which represents monotonic preferences for goods or money; it violates the continuity axiom (CON).

[20] Harsanyi (1953) in some sense anticipates Rawls's original position. This article, however, focusses on the link between cardinal utility in welfare economics and in the theory of choice under risk – not on distribution norms.

include its low information requirements: The least-favoured group of society
and their preferences are easier to determine than everybody's preferences,
as it would be required in order to calculate average utility. Related to this
is a better verifiability and, therefore, suitability as a public principle than
a utilitarian rule. Also, Rawls points at the comparatively weak assumptions
on the players' ability to commit to the bargain of the original position. In
addition, there is an advantage in terms of social stability: Inequality in the
distribution of natural assets and life-expectations "at first sight ... threat-
ens the relations between free and equal moral persons" (Rawls 1974, p. 145).
However, by using the maximin criterion "inequalities are to everyone's advantage and those able to gain from their good fortune do so in ways agreeable
to those less favored."

4.3 Binmore's Theory of the Social Contract

In his two-volume investigation *Game Theory and the Social Contract* (1994,
1998b)[21] Ken Binmore sets out "to put the case for approaching social contract questions from a game-theoretic perspective" (Binmore 1998b, p. 1).
He investigates morals, distributional justice, and fairness using Rawls's
metaphors, in particular the original position. Binmore takes a naturalistic
stance in the tradition of Thomas Hobbes and David Hume, which denies the
existence of any moral absolutes of the good or the right which are not subject to re-definitions by social convention. Commitment to individually sub-
optimal behaviour, e. g. to cooperating in a single-shot Prisoner's Dilemma, is
considered impossible. Therefore, he rejects Rawls's (1971) Kantian assumption that members of a society – who have agreed in a fair original position
on some social contract – will actually implement what is not in their best
interest *given* their places in society. Thus, only self-policing social contracts
and corresponding distribution norms are considered relevant.

Endeavouring to carry out his philosophical analysis with game-theoretic
tools, Binmore refers to the *game of life* Γ as the metaphor for an instance of
the regular interactions between people within a particular context. Its rules
are determined by physical, biological, technical, and other context-specific
constraints outside the strategic influence of individual players. The game
of life is played in an indefinitely repeated manner, defining a super-game
which will be denoted by Γ^∞. Folk theorems (cf. e. g. Osborne and Rubinstein 1994, ch. 8) imply that Γ^∞ has generically a multitude of equilibria,
many of them efficient. The (subgame perfect) Nash equilibria of Γ^∞ correspond to self-enforcing social contracts in their simplest form – for example,
they specify when players cooperate or defect in some repeated Prisoner's
Dilemma situation, or if they drive on the left or the right side of the road.

---

[21] See e. g. Holler (1996, 2000), Napel (1999a), and Sugden (2001) for critical re-
views. Linster (2000) gives a particularly sympathetic summary.

While *sustaining* an equilibrium is just a matter of the players' correctly implementing their strategies, a considerable difficulty lies in *selecting* one of the equilibria, preferably from the subset of efficient equilibria.

The selection task can be modelled as a bargaining problem $\langle U, u^D \rangle$, where U is the set of all feasible (average) payoff combinations $u = \pi(o)$ in Γ^∞ given players' personal utility functions π_i and possible outcomes $o \in O$, and where u^D describes what players can at minimum achieve without cooperation. It can be resolved by full-blown face-to-face negotiation as discussed in Chap. 1. Bargaining power – as implied by the terms of the status quo, players' attitudes towards risk and delay, outside options, breakdown risk, and sequencing of players' decisions – will determine a particular equilibrium of Γ^∞ to be played by society. However, making such deals in many different contexts may generally consume time and resources, so that there is a role for a fairness norm. Selection problems in a given context can then easily be solved by applying a rule of thumb saying: 'This is a fair deal.' The evolutionary advantage thus provided to modern man's prehistoric, hunting-and-gathering ancestors is from Binmore's naturalistic perspective the only reason why fairness and related concepts of justice and morals exist.

His approach to answering how a fairness norm is determined involves hypothetical negotiations in Rawls's original position. It is according to Binmore (1998a, p. 277) "a stylized version of a principle that we ... unconsciously apply every day." Once a society has discovered this principle or tool, and experienced the benefits of its application as an equilibrium selection algorithm, Binmore envisions that people will stop explicitly playing the physically binding repeated game of life, Γ^∞ (cf. Binmore 1998b, pp. 422ff). Rather, they will conjure up and play an almost identical game, $\tilde{\Gamma}^\infty$, which only differs from Γ^∞ in that it allows each player after each iteration of the single-shot game Γ to take an action a^R (appeal to the original position) which is not available in Γ^∞. If this action is not chosen by a given equilibrium strategy profile $\tilde{\sigma}^*$ of $\tilde{\Gamma}^\infty$, i.e. each player finds it optimal not to play a^R given that the other players stick to $\tilde{\sigma}^*_{-i}$, then $\tilde{\sigma}^*$ also specifies an equilibrium of Γ^∞. The reverse is not true, since strategies chosen in an equilibrium σ^* of Γ^∞ can lose their optimality given that the new action a^R is available. The a^R-free equilibria of $\tilde{\Gamma}^\infty$ therefore select equilibria of Γ^∞.

Binmore calls $\tilde{\Gamma}^\infty$ the *game of morals*. Member's of society accept its rules and play it because of its usefulness as a selection device – not because of any religious or deeper moral principles. Action a^R denotes the metaphorical lowering of a Rawlsian veil of ignorance by some player, and all players then have to (re-)decide – by unanimous agreement – which of $\tilde{\Gamma}^\infty$'s many equilibria shall be played from now on.[22]

---

[22] $\tilde{\Gamma}^\infty$ is a repeated game of the same type as Γ^∞. An (unmodelled) renegotiation of what equilibrium is to be played makes sense because $\tilde{\Gamma}^\infty$'s subgame starting in an arbitrary period t is equivalent to $\tilde{\Gamma}^\infty$ itself.

This setting can be used to define a *fair social contract* as (Binmore 1998b, p. 424)

> an equilibrium of the Game of Life $[\Gamma^\infty]$ that calls for the use of strategies which, if used in the Game of Morals $[\tilde{\Gamma}^\infty]$, would never leave a player with an incentive to exercise his right of appeal to the device of the original position $[a^R]$.

Thus, Binmore restates Rawls's theory of justice as fairness with the crucial difference that agreements reached in the original position are binding to no one, and will be adhered to if and only if it is in each and every individual's interest to do so.

A second important difference to Rawls is Binmore's use of a traditional Bayesian decision framework and bargaining theory. Binmore follows Harsanyi's (1977, ch. 4) model of *inter*personal comparison via *intra*personal comparison of utility. Namely, in the negotiation of the fair social contract, every player $i \in I$ uses *empathetic preferences*[23] and subjective probabilities attached to ending up in any given role of society. This assumes that each player $i \in I$, first, has the ability to mentally put himself in the position of every player $j \in I$ (trivially including i) and, second, possesses an *empathetic preference ordering* $\succsim^i$ which ranks all personality-and-outcome combinations $(j, o) \in I \times O$. For example, a set of players $I = \{1, 2, 3\}$ could correspond to an economics student, Bill Gates, and George W. Bush; the set of relevant outcomes, $O = \{o_1, o_2, \ldots, o_n\}$, could include $o_1 = \{$be chairman of Microsoft$\}$ and $o_2 = \{$be president of the U.S.A.$\}$. Then, player 1's empathetic preferences could be such that $(3, o_2) \succ^1 (2, o_1)$, i.e. the student – making gedankenexperiments – strictly prefers being George W. Bush governing the U.S.A. to being Bill Gates governing Microsoft. For all players $i \in I$, Harsanyi and Binmore make several quite demanding assumptions concerning $\succsim^i$:

Von Neumann-Morgenstern utility (VNM):
Ordering $\succsim^i$ satisfies the von Neumann-Morgenstern axioms. Hence, it has an expected utility representation v_i such that $(\forall j, k \in I)(\forall o, o' \in O)$: $\{(j, o) \succsim^i (k, o') \iff v_i(j, o) \geq v_i(k, o')\}$.

Player i's empathetic preferences $\succsim^i$ over personality-and-outcome combinations are to be distinguished from his personal preferences $\succsim_i$ over outcomes, which are represented by i's expected utility function π_i. However, a strong form of consistency between the two, and also v_i and other players' personal preferences, is required:

Perfect identification (PI):[24]
$(\forall j \in I)(\forall o, o' \in O)$: $\{v_i(j, o) \geq v_i(j, o') \iff \pi_j(o) \geq \pi_j(o')\}$.

---

[23] Harsanyi calls them *extended preferences*. A similar framework is also investigated by Sen (1970, ch. 9*).

[24] (PI) is called *principle of acceptance* by Harsanyi (1977, p. 52). This term stresses that each individual's own preferences should be accepted as the criterion for assessing his utility. Later, however, Harsanyi qualifies this and legitimizes "cor-

This means that each player $i \in I$ identifies perfectly with each member j of society in the sense that i prefers being president of Microsoft in the role of Bill Gates to being president of the U.S.A. in the role of Bill Gates if and only if Bill Gates actually prefers so. This extends to lotteries over outcomes. Finally, the existence of two outcomes $o^B, o^W \in O$ is assumed such that all members of society agree in personally considering o^B the best outcome in O, and o^W the worst outcome in O – or formally:

Best and worst outcome (BWO):
$$(\exists o^B, o^W \in O)(\forall i \in I)(\forall o \in O): \{\pi_i(o^B) \geq \pi_i(o) \wedge \pi_i(o^W) \leq \pi_i(o)\}.$$

For simplicity, assume in the following that $I = \{1, 2\}$. Player i's personal utility function π_i is only determined up to a strictly increasing affine transformation. In particular, with (BWO) in mind, it can be scaled such that

$$\pi_i(o^W) = 0 \text{ and } \pi_i(o^B) = 1. \tag{4.2}$$

By (VNM) and (PI), considering a fixed player $i \in I$, v_i for each $j \in I$ satisfies

$$v_i(j, o) = c_{j1}\pi_j(o) + c_{j2} \tag{4.3}$$

for constants $c_{j1} \in \mathbb{R}_{++}$ and $c_{j2} \in \mathbb{R}$. By (VNM), v_i can without loss of generality be scaled for all $i \in I$ such that

$$v_i(1, o^W) = 0 \text{ and } v_i(2, o^B) = 1.$$

One can then define two constants $a_i, b_i \in \mathbb{R}$ for each player $i \in I$ such that

$$a_i = 1 - v_i(2, o^W) \text{ and } b_i = v_i(1, o^B). \tag{4.4}$$

For example, $a_i = b_i = 1$ formalizes that player i is indifferent between being player 1 or 2 in either the best or the worst of outcomes. Substituting (4.2) and (4.4) into (4.3), it follows directly that $c_{12} = 0$ and $c_{22} = 1 - a_i$; this yields $c_{11} = b_i$ and $c_{21} = a_i$. Hence, each player i's empathetic utility function v_i is completely determined by the two coefficients a_i and b_i and the following relationship with the personal preferences of the two agents $j \in I$:

$$\begin{aligned} v_i(1, o) &= b_i\pi_1(o) \\ v_i(2, o) &= a_i\pi_2(o) + 1 - a_i. \end{aligned} \tag{4.5}$$

For $a_i = b_i = 1$, player i considers it better being 1 rather than 2 if and only if $\pi_1(o) \geq \pi_2(o)$.

Thus far, v_i has only been used in order to evaluate deterministic personality-and-outcome combinations. Being a von Neumann-Morgenstern utility function, its use for the comparison of lotteries is straightforward. The latter is needed to analyse players' rational behaviour in the original position.

rections" for preferences based on factual errors, and possibly even "censorship" if j's preferences conflict with "fundamental value judgements" (pp. 61f).

Binmore assumes that players attach subjective probabilities to the distinct social positions. A straightforward assumption in the considered 2-player society, I, is that roles 1 and 2 are equiprobable.[25] Then, player i's objective in the original position is to maximize

$$\Upsilon_i(o) = \frac{1}{2}b_i\pi_1(o) + \frac{1}{2}(a_i\pi_2(o) + 1 - a_i),$$

which is equivalent to maximizing

$$\tilde{\Upsilon}_i(o) = b_i\pi_1(o) + a_i\pi_2(o). \tag{4.6}$$

Υ_i can for $i \in I$ be used to formally define the bargaining problem $\langle \tilde{U}, \tilde{u}^D \rangle \in \mathcal{B}$ faced by the players in the original position or *behind* the veil of ignorance.[26] Each player, by (4.6), would then bargain as if he was trying to obtain the weighted utilitarian solution $F^{U(b_i, a_i)}(U, u^D)$ for the real bargaining problem $\langle U, u^D \rangle \in \mathcal{B}$ *in front of* the veil of ignorance.

In the original position, player i thus considers allocating a_i incremental utility units to player 1 equally good as allocating b_i incremental units to player 2. This is an important result. Namely, the ratio a_i/b_i defines a standard for *intra*personal comparisons by player i. If all players happen to have the same weights, i.e. $a_1 = a_2 = a$ and $b_1 = b_2 = b$, then consistent *inter*personal comparisons can be conducted and ratio a/b describes the social standard of comparison. Given such an identity of all players' empathetic preferences in the original position, the utilitarian solution for real bargaining problems $\langle U, u^D \rangle$ must be agreed on as society's distribution norm.[27] This has already been observed in the discussion of Rawls's theory (p. 135).

Harsanyi (1977, p. 60) expresses optimism that a common standard of comparison can be assumed, based on the great similarity among people from a common culture. But in a naturalistic theory, expected utility functions Υ_i should a priori be allowed to differ for players $i = 1$ and 2. If they differ, a conflict of interest arises in the original position. The bargaining result will – assuming rational agents – reflect the distribution of power in the original position, as described by $\langle \tilde{U}, \tilde{u}^D \rangle$. The agreed empathetic utility levels are approximated by the symmetric *Nash bargaining solution*, $F^N(\tilde{U}, \tilde{u}^D)$.[28]

---

[25] Cf. Binmore (1998b, p. 234) for the case when different subjective probabilities are attached to the two positions.

[26] In analogy to the definition of $\langle U, u^D \rangle$, $\tilde{U}$ is the set of all feasible empathetic utility combinations $u = \Upsilon(o)$ in $\tilde{\Gamma}^\infty$ given players' empathetic utility functions Υ_i and possible outcomes $o \in O$. $\tilde{u}^D$ describes in terms of empathetic utility what players can at minimum achieve without agreement.

[27] The conclusion does not yet take account of Binmore's constraints on which deals are implementable given players' realized empathetic preferences in front of the veil of ignorance.

[28] This assumes that other differences between the players such as time preferences, outside options, etc. can be neglected.

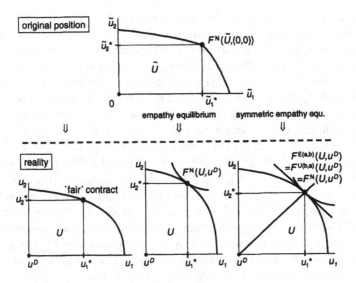

Fig. 4.3. Agreement $(\tilde{u}_1^*, \tilde{u}_2^*)$ in the original position and fair personal payoffs (u_1^*, u_2^*)

It can be argued that the coefficients of players' actual empathetic preferences, (a_1, b_1) and (a_2, b_2), will in the medium run be such that no player has an incentive to pretend having different coefficients $(\tilde{a}_i, \tilde{b}_i)$ when negotiating in the original position. A possible reason for this is that effective deception is costly. An (un-modelled) evolutionary process would then over time replace (a_i, b_i)-agents by agents who in fact have the coefficients $(\tilde{a}_i, \tilde{b}_i)$. Assuming that players can freely choose the reported empathy coefficients $(\tilde{a}_i, \tilde{b}_i) \in \mathbb{R}^2$ in a pre-negotiation stage given their real (a_i, b_i), this amounts to requiring $((a_1, b_1), (a_2, b_2))$ to be a Nash equilibrium in the original position. Binmore (1998b, p. 224) calls such a coefficient profile an *empathy equilibrium*.

The assumption that $((a_1, b_1), (a_2, b_2))$ is an empathy equilibrium has an important implication for the Nash bargain which is reached by the players in the original position (cf. Binmore 1998b, pp. 243–247, for the derivations). Namely, $F^N(\tilde{U}, \tilde{u}^D)$ implements utility levels (u_1^*, u_2^*) in front of the veil of ignorance which match exactly those given by the Nash solution $F^N(U, u^D)$ of the real bargaining problem $\langle U, u^D \rangle \in B$, where u^D corresponds to the (possibly inefficient) current play of Γ^∞ (cf. Fig. 4.3, middle). In other words, if players' empathy coefficients in the medium run evolve such that they form an empathy equilibrium, then the resulting objective functions behind the veil of ignorance guarantee that it is precisely agreed to be fair, what corresponds to the balance of power in front of it.

The result gets even sharper if social evolution of empathy coefficients is not only assumed to produce a Nash equilibrium, but a neutrally stable or even evolutionary stable strategy profile (cf. the appendix). Namely, this implies a *symmetric empathy equilibrium*, i. e. $(a_1, b_1) = (a_2, b_2)$.[29] Thus, Harsanyi's assumption of a common standard of interpersonal comparison, a/b, may be supported by naturalistic arguments. A common standard ensures that there is no conflict of interest in the original position – all players want to maximize the same convex combination of the personal utility functions π_1 and π_2. In this case, the Nash solution of the bargaining problem behind the veil of ignorance, $\langle \tilde{U}, \tilde{u}^D \rangle$, falls together with the Nash solution *and* the weighted utilitarian solution of $\langle U, u^D \rangle$ (cf. Fig. 4.3, right).

As seen in Sect. 1.2.2, if a bargaining problem $\langle U, u^D \rangle \in B$ and $(a, b) \in \mathbb{R}^2_+$ are such that two of $F^N(U, u^D)$, $F^{E(a,b)}(U, u^D)$, and $F^{U(b,a)}(U, u^D)$ coincide, then all three coincide (p. 25). If U is u^D-comprehensive, i. e. U contains every u for which $(\exists \tilde{u} \in U): u^D \leq u \leq \tilde{u}$, and the individually rational part of its Pareto frontier, i. e. $\{u \in P(U): u \geq u^D\}$, contains no line segment which is parallel to the u_1- or u_2-axis, then U is called *strictly u^D-comprehensive*. For bargaining problems with strictly u^D-comprehensive U, the weighted egalitarian and the Rawlsian solution coincide. Then, granted that players' are Bayesian maximizers in the original position and society is in a symmetric empathy equilibrium, altogether four contenders for the model society's norm of distributional justice investigated in Sects. 4.1 and 4.2 can be deduced as the collective decision rule.

A symmetric empathy equilibrium may take considerable time to be reached. Binmore considers it only a medium-run phenomenon. However, it is important that *in the medium run* players' standards of intrapersonal comparison adjust to a society-wide standard of interpersonal comparison which equates Rawlsian,[30] weighted egalitarian, proportional, and Nash solution of the bargaining problem $\langle U, u^D \rangle$ defined by Γ^∞ in front of the veil of ignorance.

In the short run, some standard a/b of interpersonal comparison which need not reflect power in Γ^∞ can be in place.[31] For example, the game of life Γ may have been exogenously changed to a slightly different game Γ'. Binmore asks which short-run distribution rule must be used in front of the veil of ignorance in order to move quickly to an efficient equilibrium of the new repeated game of life, Γ'^∞, subject to not inducing any player to choose a^R in the associated game of morals, $\tilde{\Gamma}'^\infty$. This calls for a distribution norm which

---

[29] Since only the ratios a_i/b_i turn out to influence the deal reached in the original position and hence payoffs from deception, evolutionary forces can only select for these ratios. More precisely, a 'symmetric' empathy equilibrium is thus characterized by $a_1/b_1 = a_2/b_2$, rather than $(a_1, b_1) = (a_2, b_2)$.

[30] This assumes strict comprehensiveness.

[31] Binmore does not specify any dynamic adaptation process which moves players' intrapersonal standards away from an obsolete common standard a/b to a new symmetric empathy equilibrium a'/b'. The issue of signalling false coefficients $(\tilde{a}_i, \tilde{b}_i)$ if a/b is no empathy equilibrium is unfortunately not explicitly taken up again.

implements in real life what would result from Nash bargaining behind the veil of ignorance if $((a, b), (a, b))$ is no empathy equilibrium (but symmetric).

In negotiating to play an efficient equilibrium of Γ'^∞ in the original position, the old contract agreed for Γ^∞ is taken to be the new status quo, u^D, i.e. all players keep the corresponding utility levels in the worst case of disagreement in the original position. By definition, this new status quo is egalitarian given the unchanged ratio a/b used in its original derivation, i.e.

$$bu_1^D = au_2^D + 1 - a$$

If $\tilde{u}^D$ is scaled to be $(0, 0)$ in the original position, this fixes $a = 1$ in players' empathetic preferences.

In Binmore's setting, the players are constrained to original-position agreements which are self-policing after the veil of ignorance is lifted. In particular, none of the players must prefer to re-enter the original position with the hope that a new toss of the imagined coin which allocates social roles improves his realized empathetic utility level. For example, an agreement that one of the players becomes slave to the other may be desirable for both players in terms of expected empathetic utility. It is no feasible agreement, though, since the actual slave will want to throw the social dices again. This restricts the set U of possible agreements to social contracts which specify the same realized empathetic utility for player i no matter what random configuration of social roles is realized, i.e. the agreed outcome o^* must satisfy $v_i(1, o^*) = v_i(2, o^*)$.[32]

Let $\tilde{U}_{12}$ denote all empathetic utility combinations $(\tilde{u}_1, \tilde{u}_2)$ such that $\tilde{u}_1 = v_1(1, o) = b_1\pi_1(o)$ and $\tilde{u}_2 = v_2(2, o) = a_2\pi_2(o) + 1 - a_2$ for some outcome $o \in O$ (cf. Fig. 4.4). This is the set of empathetic utility combinations feasible in the original position conditioned on player 1 (2) in fact turning out to be player 1 (2) in front of the veil of ignorance. Similarly, let $\tilde{U}_{21}$ denote all empathetic utility combinations $(\tilde{u}_1, \tilde{u}_2)$ such that $\tilde{u}_1 = v_1(2, o) = a_1\pi_2(o) + 1 - a_1$ and $\tilde{u}_2 = v_2(1, o) = b_2\pi_1(o)$ for some outcome $o \in O$. This conditions feasible payoff combinations in the original position on player 1 (2) turning out to be player 2 (1) after the veil is lifted. The restriction to those deals which are invariant, in terms of realized empathetic utility, to repeated tossing of the coin is then captured by specifying $\tilde{U} = \tilde{U}_{12} \cap \tilde{U}_{21}$ as the set of feasible social contracts in the original position. Exploiting that some standard a/b of interpersonal utility comparison is given and that $\tilde{u}^D$ is normalized to $(0, 0)$, players have to resolve a *symmetric* bargaining problem $\langle \tilde{U}, (0, 0) \rangle \in B$ in the original position. The Nash solution $F^N(\tilde{U}, (0, 0))$ then specifies symmetric efficient empathetic utility levels $\tilde{u}_1^* = \tilde{u}_2^*$ independent of the social role allocation. This implies

$$bu_1 = au_2 + 1 - a$$

---

[32] The agreed contract need not – though this is suggested by the notation o^* – be deterministic. It may even be a *contingent contract* which specifies different outcomes for different realizations of social positions.

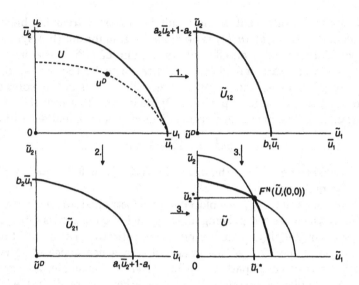

Fig. 4.4. Deriving the Nash solution of the bargaining problem in the original position (using $a_1 = a_2 = 1$, $b_1 = b_2 = b$ in the third step)

in terms of the associated personal utility levels in the changed repeated game of life Γ'^∞. Recalling that the normalization $\tilde{u}^D = (0,0)$ entails $a = 1$, this is equivalent to

$$u_2 = bu_1,$$

i.e., additional gains from cooperation which are available in the changed game of life Γ' are shared *proportionally*. Thus, the only norm of distributional justice which implements in front of the veil of ignorance what rational bargainers would agree on behind the veil of ignorance given ratio a/b is the proportional or weighted egalitarian bargaining solution $F^{E(a,b)}$.

Assuming that the bargaining problem $\langle U, u^D \rangle$ modelling equilibrium selection in front of the veil of ignorance is strictly u^D-comprehensive, $F^{E(a,b)}$ is equivalent to the Rawlsian solution F^R applied to comparable utility gains. Generally, F^R specifies never less – and possibly more – utility for each player than $F^{E(a,b)}$. Hence, Rawls's maximin equity criterion is a stable social norm if the disadvantaged player, before the possible play of a^R, anticipates both a bargaining outcome implementing $F^{E(a,b)}$ *and* himself ending up as the disadvantaged player again.[33]

---

[33] Binmore (1998b, p. 439) recalls that a social contract is defined to be fair "if nobody playing the Game of Morals would have an incentive to appeal to the original position." His claim that "all social contracts satisfying the maximin criterion have this property" is not entirely consistent with having earlier ruled out utilitarian contracts on the grounds that the disadvantaged player will want a new coin toss to determine who has what position. He seems to require coin-

Holding that all relevant social coordination problems correspond to strictly comprehensive sets of feasible payoffs, Binmore vindicates Rawls's conclusion that the maximin equity criterion will be agreed on by free and rational bargainers in the original position. In his own words, Binmore (1994, p. 53) manages "to pull a maximin rabbit out of a hat into which only Bayesian decision theory is placed." The reconstruction of Rawls's maximin conclusion via the proportional bargaining solution is accomplished by restricting attention to contracts with a particular stability anticipated by Rawls (1974, p. 144):

> ... in the original position the parties are to favor those principles compliance with which should prove more tolerable, whatever their situation in society turns out to be.

Namely, there must be no incentive for a renegotiation of the social contract after the veil of ignorance is lifted, and possibly unpleasant social positions have been assigned.

In Binmore's theory, Rawls's maximin criterion is to be applied to appropriately re-scaled gains from social cooperation. The scaling factors correspond to the worthiness coefficients used by the weighted egalitarian solution. It is important that Binmore's theory specifies how these coefficients of social worthiness are determined: In the medium run, they implement precisely the distribution resulting from the factual balance of power in Γ^∞ as the fair distribution. Hence, any moral content of the concepts of 'fairness' and 'social justice' is lost. The social evolution of values, according to Binmore, makes people accept as just what they do not have the power to change. Still, adhering to (evolving) principles of justice which are determined by a self-imposed game of morals, $\tilde{\Gamma}^\infty$, is mutually beneficial because it facilitates equilibrium selection in the actual game of life, Γ^∞.

4.4 Discussion

Several competing distribution norms are available to a society which itself creates its own rules for mutual cooperation. These rules give a new normative interpretation to bargaining solutions, which have been introduced in Sect. 1.2 to predict, not to prescribe agreements. Section 4.1 has investigated the properties of prominent solutions under this interpretation. Sections 4.2 and 4.3 have tried to identify which rule would be chosen if everyone forgot his current position in society, and which rule should be the prevalent one on evolutionary grounds.

Binmore draws somewhat unpleasant conclusions concerning the latter question. He predicts a proportional norm whose weights are in the medium run determined by power – not any first moral principles. Binmore (1998b,

---

tossing stability only before players re-emerge from the original position, not in the game of morals.

p. 459, fn. 48) correctly asks: "... is it reasonable to base one's reactions to a claim about how the world works on whether one likes or dislikes its consequences?" Today, it is difficult to hold that no absolute good or right obliges mankind to condemn, for example, child abuse or slavery. Still, the evidence seems to be on the side of Binmore's moral relativism – paedophilia was accepted not only in ancient Greece or Rome, and it is not that long ago that a civil war was necessary to abolish slavery in North America. Binmore himself points at Rawls's *two* principles of justice (see p. 136), of which he only has investigated and deduced the second, part (a). Qualifying his weighted egalitarian conclusion by pointing at first principles concerning "the most extensive basic liberty" is, however, problematic in a naturalistic framework.

It is open to discussion if Binmore presents a naturalistic theory at all. Sugden (2001) takes a particularly critical stance. Although Binmore is careful to provide anthropological evidence for several of his claims (cf. Binmore 1998a, pp. 285–298), Sugden scathes Binmore for imposing (VNM) and (PI) on players' empathetic preferences, for using expected utility theory in the first place, and for referring to Rubinstein's SPE analysis in order to justify the Nash solution. Even if all this may be admitted as a first approximation, too much of Binmore's analysis, in the author's view, depends on Rawls's original position. This is a hypothetical state; it is no more – and no less – than a fascinating theoretical construct. It elaborately captures the aspect of impartiality which is central to the discussion of justice and fairness in reality, too. However, rational bargaining behind a veil of ignorance is clearly not a feature of reality, and not a natural feature of a naturalistic explanation.

Turning back to Rawls's original conclusion that the maximin equity criterion is the fair one, some of its controversial implications need to be pointed out. In the extreme, it requires all social surplus to be dedicated to the improvement of the plight of the most disadvantaged members of society – no matter how small the feasible improvement is, and how much greater improvements for other groups of society would be feasible using the same resources. Harsanyi (1975, p. 596) gives the example of a basically healthy patient A and a terminal cancer victim B who are both critically ill with pneumonia; there is only sufficient antibiotic to treat one of them. He argues that maximin ethics would call for B to be treated, while a utilitarian doctor would save patient A. Also, Harsanyi criticizes that if an education budget had to be allocated either to education in higher mathematics for talented students or remedial training for retarded persons, Rawls's difference principle would imply the latter, while a utilitarian "would accomplish 'much more good,' and would create a much deeper and much more intensive human satisfaction" (p. 596) by choosing the former. Rawls (1974, p. 142) replies that

> ... the maximin criterion is not meant to apply to small-scale situations, say, to how a doctor should treat his patients or a university its students.
> ... Maximin is a macro not a micro principle.

He probably has a point in that Harsanyi's examples are somewhat artificial. They make the utilitarian judgement the more commonsensical in view of implicit dynamic aspects of the situation.[34] However, it seems odd from a normative point of view that people should follow different ethics for micro and macro questions. Though it would be possible to e. g. design a constitution in a Rawlsian way and in this framework have everybody act as a rule utilitarian, it would be surprising if a consistent theory of justice were to yield this prescription.[35]

Binmore argues that relevant social bargaining problems satisfy the comprehensiveness requirement which makes the egalitarian and the Rawlsian bargaining solutions equivalent in terms of worthiness-adjusted incremental utility. Therefore, he does not focus too much on the different implications of his and Rawls's conclusion. Given that he sees worthiness to be determined by the realities of power, people's intuitive responses to above questions of remedial training for retarded persons and pneumonia treatment for a terminally ill patient are considerably better in line with Binmore's proportionality principle than Rawls's difference principle.

An argument in favour of the proportional solution not mentioned above is that it is the only monotonic and decomposable solution which is individually rational and, for comprehensive problems, weakly Pareto efficient (cf. Sect. 1.2.2 and Kalai 1977b). This makes it the best distribution norm in a growing economy, and a society which can implement reforms only in several small steps: In the first case, possibly asymmetric growth of players' opportunities is ensured to be to everybody's advantage. Similarly, the step-by-step approximation of an agreed efficient distribution entails mutual benefit at every stage if partial gains are allocated proportionally. And importantly, the partial realization of gains will not suddenly render a different than the agreed-upon efficient distribution the fair one.

This chapter has discussed rather philosophical questions in terms of game-theoretic models of bargaining. As in Binmore (1994, 1998b), formal statements have been confined to the case of a 2-person society. This case is highly artificial, even if one thinks of it in terms of two homogeneous classes of society. Although the investigated bargaining solutions can readily be generalized to an n-person setting, this assumes that only general contracts unanimously agreed by all n players are feasible. If sub-groups of the players can form coalitions and agree on partial contracts, the matter becomes considerably more complex. A formal treatment of negotiations behind the veil of ignorance is then even more reduced to serve illustrative purposes than it has been the case in Binmore's analysis.[36] The balance of power in

---

[34] For example, patient A may later discover more economical antibiotics or a cure for cancer; mathematic algorithms may improve aggregate productivity and allow for much more remedial training *and* education in higher maths than is presently affordable.

[35] From a positive point of view, it seems likely that people *do* follow different principles on different levels.

an n-player society with endogenous coalition formation depends highly on the bargaining procedure, which also should emerge endogenously in a truly naturalistic theory. This indicates that a comprehensive model of the evolution of a contractarian conception of justice is presently beyond the reach of bargaining theory.

Binmore assumes complete information not only about empathetic preferences in the original position but also personal preferences in the recurring game of life. Then, taking bargaining theory literally, there is no social benefit from having a fairness norm: Both Nash's and Rubinstein's models of bargaining predict efficient agreement with neither delay nor waste of surplus. The complexities of modelling bargaining under incomplete information have been indicated in Sect. 1.3.3. They make clear-cut conclusions either impossible in view of the multiplicity of equilibria which arises frequently, or arbitrary if particular of the many available equilibrium refinements were applied.

A similar shortcoming is Binmore's statement of an essentially dynamic theory – with crucial distinctions between the short and the medium run – via entirely *static models*. There is no link between the short-run use of a proportional distribution norm and a supposed medium-run evolution towards symmetric empathetic equilibria. The notion that an evolutionary process shapes players' empathy coefficients is already elusive, since their fitness is supposedly evaluated in the original position, not the real world (see Binmore 1998b, pp. 224ff). But global convergence to some evolutionary stable state, in which players' usually divergent standards of intrapersonal comparison coalesce, is no more than speculation. Unfortunately, a symmetric empathy equilibrium is of critical importance to Binmore's theory.

Although the evolution of social norms and ethical principles is too complex to be tackled satisfactorily by bargaining theory and other game-theoretic tools, the author supports Binmore's "case for approaching social contract questions from a game-theoretic perspective" and congratulates him on his achievement. He is correct in replying to Sugden: "I offer some foundations where my predecessors offer none" (Binmore 2001, p. F244). The game-theoretic approach uses well-specified behavioural assumptions which are less controversial than the maximin choice rule assumed by Rawls. In addition, testable predictions – in Binmore's theory, the relationship between worthiness and bargaining power[37] – are generated. This facilitates a pos-

---

[36] Compare Binmore's (1998b, pp. 495–499) own account of his theory's – probably unavoidable – technical shortcomings.

[37] Binmore (1998b, Sect. 4.7) investigates the comparative statics of 'worthiness' implied by the comparative statics of the Nash bargaining solution. For example, worthiness is increased by *need* if the latter is interpreted as willingness to take risks in order to obtain a desired object – be it food or a new Porsche. Similarly, the necessity to exert *effort* before the consumption of social goods, reflected by a reduction of the player's total payoff from cooperation, reduces a player's worthiness (pp. 464–466). High *social standing*, interpreted as a great status

itive approach to traditionally normative questions of fairness and justice. Moreover, the more precisely a theory is formulated, the more evident are its problematic assumptions or flaws. This makes discussion more efficient (though perhaps less creative).

Binmore's most controversial assumption is – in the author's view – Rawls's original position. In a naturalistic theory, the surplus-sharing convention should evolve within the game of life, not from hypothetical bargaining behind an appropriately thick veil of ignorance. Concerning this point, Young's (1993b, 1998, ch. 9) bargaining and social contract models and also the model of adapting aspirations investigated in Chap. 2 have greater naturalistic appeal.

Young's models support either the Nash solution or the Kalai-Smorodinsky solution as the stable social norm. The model of aspiration-based bargaining developed in Chap. 2 does not select a particular bargaining solution; the introspective satisficing heuristic allows that – alternating – either the egalitarian (or maximin) payoff combination or the power-based asymmetric one satisfies both players or, one may say, is regarded as just. Both adaptive play and the aspiration heuristic do not distinguish between players' response to, first, an efficient distribution which does not give them what they expect and, second, inefficient play. In reality, the former tends to produce different reactions from the latter – evoking calls for others to adapt, rather than causing a revision of one's own behaviour. Capturing this aspect of fairness may be a strength of Binmore's game of morals.

In this chapter, bilateral bargaining theory has been applied to deal with a more fundamental question than the division of a euro or terms of bilateral exchange. It may be obvious, but many constitutional arrangements and political institutions implement particular solutions to the bargaining situations typically faced by any association of free and rational individuals. These solutions correspond to different concepts of social justice and fairness. The need to have such concepts, in fact, fundamentally depends on the existence of bargaining problems. As indicated by the discussion of experimental evidence on bargaining in Sect. 1.5, knowledge about people's concerns for distributional justice and fairness can improve game theorists' understanding of bargaining. This chapter claims that the reverse is also true.

---

quo utility, u_i^D, similarly decreases worthiness (p. 468), while increased *ability*, defined as the capacity to produce the same output with less effort, increases it.

Appendix:
Game-theoretic Concepts, Notation, and Results

This appendix collects game-theoretic concepts, notation, and results which are used in the main text. It focusses on non-cooperative theory and bargaining. Several excellent textbooks provide more detailed information; see e. g. Fudenberg and Tirole (1991), Binmore (1992), or Osborne and Rubinstein (1994). Owen (1995) is a particularly useful reference for cooperative game theory and power indices. Weibull (1995) and Young (1998) are highly recommended for evolutionary game theory.

Let I be the set of *players*, representing economic or social agents such as individuals, firms, governments, etc. For one given player $i \in I$, the other player – or all other players, if more than two are considered – will be denoted by $-i$. Unless stated otherwise, $I = \{1, 2\}$. In a non-cooperative game, each player $i \in I$ can choose a deterministic *pure strategy* s_i from the set S_i. The players may also independently randomize over pure strategies. This yields *mixed strategies*, σ_i, which are elements of the space $\Delta(S_i)$ of probability distributions over S_i.

Each *pure strategy profile* or *configuration* $s = (s_1, s_2) \in S_1 \times S_2$, determines a unique *outcome* $o(s_1, s_2)$ of the game. In the context of bargaining, an outcome can correspond to a particular agreement like an allocation of goods, or disagreement. Each player is assumed to have a *(personal) preference ordering* $\succsim_i$ over the set of all feasible outcomes O.[1]

More demanding, it is usually also assumed that players have complete and transitive preferences with respect to probability distributions or *lotteries* over outcomes, $l \in \Delta(O)$, which satisfy the *von Neumann-Morgenstern axioms*. The *continuity axiom (CON)* requires that for any three strictly ranked lotteries, mixtures of the best and the worst can be found such that one mixture is better and another mixture is worse than the intermediately ranked lottery, i. e. $(\forall l, l', l'' \in \Delta(O))$: $\{l \succ_i l' \succ_i l'' \implies (\exists \alpha, \beta \in (0,1))$: $\alpha l + (1 - \alpha)l'' \succ_i l' \succ_i \beta l + (1 - \beta)l''\}$. The *independence axiom (IND)* specifies that the ranking of two lotteries is preserved by analogously mixing both with a third lottery, i. e. $(\forall l, l' \in \Delta(O))$: $\{l \succ_i l' \implies (\forall l'' \in \Delta(O))(\forall \alpha \in (0,1))$: $\alpha l + (1 - \alpha)l'' \succ_i \alpha l' + (1 - \alpha)l''\}$. Under these two

---

[1] $\succsim_i$ induces the following relations: 1) $o \precsim_i o' :\Longleftrightarrow o' \succsim_i o$, 2) $o \sim_i o' :\Longleftrightarrow o \succsim_i o' \wedge o \precsim_i o'$, 3) $o \nsucc_i o' :\Longleftrightarrow \neg(o \sim_i o')$, 4) $o \succ_i o' :\Longleftrightarrow o \succsim_i o' \wedge o \nsucc_i o'$, and 5) $o \prec_i o' :\Longleftrightarrow o' \succ_i o$.

assumptions, player i's ordering $\succsim_i$ can be represented by an *expected utility function* or *payoff function* $\pi_i \colon \Delta(O) \to \mathbb{R}$ such that $\pi_i(l) = \int u_i(o)\,dl(o)$ holds for a function $u_i \colon O \to \mathbb{R}$. Function u_i specifies the utility of deterministic outcomes $o \in O$ to player i. Both u_i and π_i are uniquely defined up to an order-preserving affine transformation. Player i weakly prefers lottery l to l' if and only if $\pi_i(l) \geq \pi_i(l')$.

Each *mixed strategy profile* $\sigma = (\sigma_1, \sigma_2) \in \Delta(S_1) \times \Delta(S_2) = \Delta(S_1, S_2)$ defines a unique lottery $l \in \Delta(O)$.[2] So, payoff functions π_i are well-defined on the space of mixed strategy profiles, and one can also write $\pi_i \colon \Delta(S_1, S_2) \to \mathbb{R}$. Given its expected utility form, it suffices to specify π_i for all pure strategy profiles. $\pi_i(s_1, s_2)$ is written instead of $\pi_i(\sigma_1, \sigma_2)$ when σ_i is a degenerate mixed strategy with unit mass on s_i.

Strategy sets and players' preferences determine a set of *feasible payoff combinations* $U(\Delta(S_1, S_2), \pi) \subseteq \mathbb{R}^2$ where $(u_1, u_2) \in U(\Delta(S_1, S_2), \pi)$ if and only if there is a mixed strategy profile[3] $\sigma \in \Delta(S_1, S_2)$ for which $(\pi_1(\sigma), \pi_2(\sigma)) = (u_1, u_2)$. When strategy space and payoff functions are not of explicit interest or unknown, one simply writes U to denote the set of feasible payoff combinations.

A payoff vector $u \in U$ for which some $u' \in U$ satisfies $u' \geq u$, meaning that $u' \geq u$ and $u'_i > u_i$ for at least one player i, is said to be *Pareto-dominated* by u'. If no $u' \in U$ Pareto-dominates $u \in U$, then u is *Pareto-efficient*. The subset of Pareto-efficient elements of U is called the *Pareto frontier* of U and is denoted by $P(U)$. If U contains at least two distinct Pareto-efficient payoff combinations, it can be interpreted as the formalization of a bargaining situation: the players have the opportunity to collaborate for mutual benefit in more than one way. Note that players' individual rationality may prevent them from exploiting this opportunity if no binding cooperation agreements can be made as e. g. in the well-known *Prisoner's Dilemma game* with $S_1 = S_2 = \{C, D\}$ and $\pi_1(D, C) = \pi_2(C, D) > \pi_i(C, C) > \pi_i(D, D) > \pi_1(C, D) = \pi_2(D, C)$ for $i \in I$.

Cooperative game theory assumes that a binding agreement to select any feasible payoff combination $u \in U$ can in some unspecified way be made by the players. Additional information about U may be relevant in the context of bargaining. Often, a vector $u^D \in U$, called *disagreement point* or *impasse point* or *status quo point*, describes the players' payoffs in case they fail to reach an agreement. Payoff vectors $u^{O_1} \in U$ and $u^{O_2} \in U$ may specify utility after an endogenous breakdown of negotiations, i. e. when player 1 or 2, respectively, chooses to leave the negotiation table and to pursue an available *outside option*. If there is a risk that ongoing negotiations are exogenously stopped at random, a *breakdown point* $u^B \in U$ describes the resulting util-

---

[2] $\Delta(S_1, S_2)$ here and later abbreviates the product space of $\Delta(S_1)$ and $\Delta(S_2)$, not the space $\Delta(S_1 \times S_2)$ of all joint distributions on $S_1 \times S_2$.

[3] This profile need not be unique. One can also define $U = U(S, \pi)$ for the case of pure strategies. The convexity of U implied by randomization is, however, technically often desirable.

ity combination. *Cooperative bargaining theory* tries to identify which payoff combination(s) can be expected from the rational interaction of players in this setting.

While cooperative bargaining theory deals with a bargaining situation as described by the abstract set U and, perhaps, distinguished points in U, *non-cooperative bargaining theory* and, more generally, *non-cooperative game theory* analyse interaction which is based on explicit rules of the game. Apart from what is specified by players' strategies, no binding agreements are feasible.

In a prominent class of non-cooperative games, players' strategy sets as well as their payoff functions are *common knowledge*, i.e. both players know them, know that both know them, know that both know that both know them, etc. (ad infinitum), and the players simultaneously choose one of their respective pure or mixed strategies. These assumptions, the set of players, I, the strategy space $S = S_1 \times S_2$ or $\Delta(S_1, S_2)$, and the joint payoff function $\pi = (\pi_1, \pi_2)$ define a *game in strategic form* or *game in normal form*. Depending on whether the players must use pure strategies or can independently randomize, a strategic-form game is denoted by $\langle I, S, \pi \rangle$ or $\langle I, \Delta(S_1, S_2), \pi \rangle$, respectively.

An alternative representation of a game, introduced by von Neumann and Morgenstern (1944) and developed by Kuhn (1953), involves the explicit description of which actions can be taken by whom with what available information at each specified stage of the game. The main ingredient of such a representation is a *game tree*, which consists of a finite set[4] $\mathcal{N}$ of *nodes* and an asymmetric, transitive binary *precedence relation* $\prec$ on $\mathcal{N}$ which satisfies the arborescence property: If distinct nodes n and n' are both predecessors of node n'', then either n precedes n', or n' precedes n. Moreover, there is a unique *initial node* $n^0 \in \mathcal{N}$ which has no predecessor. Nodes which precede no other node, i.e. are without successor, are *terminal nodes* and specify *outcomes*. All terminal nodes are collected in the set $\mathcal{T}$. All non-terminal nodes are called *decision nodes* and collected in $\mathcal{D} = \mathcal{N} \setminus \mathcal{T}$. The set of *immediate successors* of a decision node $n \in \mathcal{D}$ will be denoted as $s(n)$. The set of all immediate or indirect successors of n is $S(n)$.

Given the game tree, a function $\iota \colon \mathcal{D} \to I \cup \{N\}$ specifies for every decision node the player who has to move. N is an additional player called *Nature* which is introduced in order to model exogenous chance moves. For each decision node $n \in \mathcal{D}$ there is a set of *available actions* $A(n)$ which player $\iota(n)$ can choose from. A bijective mapping $\alpha(n) \colon s(n) \to A(n)$ specifies which unique action $a \in A(n)$ has to be taken by player $\iota(n)$ in order to take the game from n to a given immediate successor node $n' \in s(n)$.

There is a partition $\mathcal{H}$ of the set of decision nodes including the singleton $\{n^0\}$, whose elements are called *information sets*. First, $\mathcal{H}$ has to be such that no successor of n belongs to $H \in \mathcal{H}$ if $n \in H$. Second, the same player must

---

[4] The infinite case is technically more involved, but, in principle, similar (cf. fn. 5).

be assigned by ι to all nodes $n \in H$, so that with slight abuse of notation $\iota(H)$ denotes the player having the move in H. Third, the same set of available actions $A(n)$ must be assigned to each node in H, so that with another slight abuse of notation $A(H)$ refers to the actions available in H. Information sets describe what a player knows about previous moves in the game: He can distinguish moves that have lead to distinct information sets H and H', but not moves that have lead to the same set H.

The collection $[I, \mathcal{N}, \prec, \iota, \{A(n)\}_{n \in \mathcal{N}}, \{\alpha(n)\}_{n \in \mathcal{N}}, \mathcal{H}]$ defines an *extensive game form*. To make it a game, players' preferences over terminal nodes and probabilities for Nature's moves must be specified. As before, von Neumann-Morgenstern utility functions π_i represent player i's preferences over terminal nodes, i.e. outcomes, or the strategy configurations which induce them. A probability distribution $\varrho(n)$ on $A(n)$ defines for each node n with $\iota(n) = N$ the exogenous likelihood of Nature 'choosing' a particular action in $A(n)$ when it is urged to 'play'. An interdependent decision situation described by all these primitives is called an *extensive game* or a *game in extensive form*. When all above primitives are common knowledge, it is a game of *complete information*. It is formally referred to by the 9-tuple $\Gamma = \langle I, \mathcal{N}, \prec, \iota, \{A(n)\}_{n \in \mathcal{N}}, \{\alpha(n)\}_{n \in \mathcal{N}}, \mathcal{H}, \pi, \{\varrho(n)\}_{\iota(n)=N} \rangle$, but often it is convenient to work with only a graphical or verbal representation. For example, Fig. 1.13 (p. 61) illustrates a bargaining game without chance moves (singleton information sets are not specially indicated) and Fig. 1.12 (p. 46) one in which player 1 does not observe Nature's move.

Unless players' lack of perfect rationality is explicitly mentioned, this book is concerned only with games of *perfect recall*, i.e. players remember what they previously knew and did at each stage. This restricts the possibilities of partitioning the game tree into information sets. If, in particular, every information set is a singleton, the game is said to be one of *perfect information*. Otherwise it is a game of *imperfect information*.

A strategic-form game may be represented as an extensive game of imperfect information in which one player, say 1, moves first, and player 2 subsequently chooses an action without knowledge of 1's choice. There is no need to distinguish actions $a_i \in A(n)$ (for $\iota(n) = i$) from strategies in this case. Typically, though, a player collects at least some information about the course of play in an extensive game. Then, it is important to distinguish actions taken at particular information sets from a player's overall strategy. Player i's *pure strategy in an extensive-form game* Γ is a complete plan of i's actions which specifies a deterministic move for every, even counter-factual contingency. Let $\mathcal{H}_i$ refer to all information sets at which i has the move, i.e. $H \in \mathcal{H}_i$ implies $\iota(H) = i$. Then, formally, a pure strategy is a function $s_i : \mathcal{H}_i \to \bigcup_{H \in \mathcal{H}_i} A(H)$ which maps each information set $H \in \mathcal{H}_i$ in which i could theoretically be called upon to act to a feasible action $s_i(H) \in A(H)$ – no matter whether H is reachable or not under the actions that s_i prescribes for preceding decision nodes. The sequence of reached nodes as the first t

actions prescribed by a strategy profile s are executed is the *history* $h^t(s)$ induced by s. Let $h(s)$ denote the *terminal history* induced by s, starting with n^0 and ending in some terminal node in a finite game.[5]

Collecting all of i's pure strategies in extensive game Γ in a strategy set S_i, a game $\langle I, S, \pi \rangle$ can be defined as Γ's strategic-form counterpart. Many of the strategies in S_i will differ only with respect to actions in information sets that can never be reached. Or they induce different outcomes but identical payoffs to player i for any fixed s_{-i}. When for all players i, all strategies s_i that are equivalent for i in payoff terms are condensed into a single strategy s_i', and the strategy sets are correspondingly reduced, one obtains the *reduced normal-form* representation $\langle I, S', \pi \rangle$ of extensive game Γ.

Randomization in extensive games can take two different forms. Players can either randomize over entire pure strategies, implying that moves by the same player at different information sets are statistically dependent. This defines a *mixed (extensive game) strategy* $\sigma_i \in \Delta(S_i)$. Alternatively, a player may randomize at each information set independently. Then, i's strategy becomes a collection, $b_i = \{b_i(H)\}_{H \in \mathcal{H}_i}$, of independent probability distributions – specifying one $b_i(H) \in \Delta(A(H))$ for each information set H in which i might have to move. Such a collection b_i is called a *behavioural strategy*. Player i's behavioural strategies are collected in the set B_i. In terms of the induced lotteries over terminal histories, mixed and behavioural strategies turn out to be equivalent for extensive games of perfect recall.[6] Therefore, either concept can be used in a discretionary way.

If some player i's preferences are not common knowledge, the game is incompletely specified from $-i$'s point of view. As demonstrated by Harsanyi (1967/68), a game of *incomplete information* about preferences can be transformed into a game of complete preference information but imperfect information about moves: One interprets i's preferences as determined by the realization of a random variable – modelled as an initial move by Nature in which it assigns a particular *type* θ_i to i.[7] θ_i is usually revealed only to player i himself, i.e. it is *private information*.

A reduced form of such a game of incomplete information can be constructed by, first, collecting all player types that have positive probability in a set $\Theta = \Theta_1 \times \Theta_2$, and stating a commonly known joint probability distribution F_Θ of types. Second, one extends the domain of each player i's pure strategies from $\mathcal{H}_i$ to $\mathcal{H}_i \times \Theta_i$, so that a strategy now contains a separate complete plan of actions for each type $\theta_i \in \Theta_i$. And, finally, one extends the domain of each

---

[5] An extensive game with infinitely many stages may not end in a terminal node. A terminal history is then possibly an infinite sequence of nodes induced by a strategy profile. Preferences can be defined on either these terminal histories or the profiles.

[6] For any behavioural strategy b_i, there is a mixed strategy σ_i which induces the same distribution over terminal histories for any mixed or behavioural strategy chosen by $-i$, and vice versa.

[7] One equivalently speaks of Nature sending the *signal* $\theta_i \in \Theta_i$ to player i.

player i's utility function, π_i, from $S_1 \times S_2$ (or $\Delta(S_1 \times S_2)$) to $S_1 \times S_2 \times \Theta_i$ (or $\Delta(S_1 \times S_2) \times \Theta_i$). The 5-tupel $\langle I, S, \Theta, F_\Theta, \pi \rangle$ or $\langle I, \Delta(S_1, S_2), \Theta, F_\Theta, \pi \rangle$ then describes a *Bayesian game*.

Above definitions and terminology apply to general economic, social, or even biological interaction. The motivation for using this abstract setting is the large tool-box available for the analysis of games. Here, *analysis of a game* refers to the attempt to select a subset of feasible outcomes or strategy configurations by certain objective criteria.

A game-theoretic *solution* is a mapping from a space of games to the space of outcomes or strategy profiles. In cooperative game theory, a solution is usually specified explicitly, e. g. by stating a function that maps each element of a space of bargaining games to a feasible payoff combination. *Set-valued solutions* are distinguished from (vector) *value solutions*. For non-cooperative games, one usually refers to a particular *equilibrium concept* which classifies each strategy profile as either an equilibrium or not. The implicit mapping of a given class of non-cooperative games to the power set of possible equilibria is a solution.

For some games, it is not necessary to consider entire strategy profiles to narrow down the set of 'plausible' or 'rational' outcomes. If a strategy $\sigma_i \in \Delta(S_i)$ performs strictly better than a strategy $\sigma_i' \in \Delta(S_i)$ no matter what is played by $-i$, i. e. $(\forall \sigma_{-i} \in \Delta(S_{-i})): \pi_i(\sigma_i, \sigma_{-i}) > \pi_i(\sigma_i', \sigma_{-i})$, then σ_i' is said to be *strictly dominated* by σ_i. Strictly dominated strategies are an impossible feature of a rational strategy profile, i. e. one based on players' true von Neumann-Morgenstern preferences.[8] One can in some games iteratively delete strictly dominated strategies from S until only a singleton profile s^* remains as the rational prediction of play. If above inequality is weak, σ_i *weakly dominates* σ_i'. Iterative deletion of weakly dominated strategies sometimes also produces a unique prediction, which may, however, depend on the order of deletion. To check for dominance it suffices to consider only pure strategy profiles s_{-i}. But a pure strategy s_i' can be strictly dominated by a mixed strategy σ_i, although it is un-dominated by all other pure strategies s_i.

The most prominent concept of non-cooperative game theory is the equilibrium notion introduced by John F. Nash (1950b, 1951). A strategy profile $\sigma^* \in \Delta(S_1, S_2)$ constitutes a (mixed strategy) *Nash equilibrium (NE)* of the normal-form game $\langle I, \Delta(S_1, S_2), \pi \rangle$ if $(\forall i \in I) (\forall \sigma_i' \in \Delta(S_i)): \pi_i(\sigma_i^*, \sigma_{-i}^*) \geq \pi_i(\sigma_i', \sigma_{-i}^*)$. In words, σ^* is a NE profile if and only if no player i has a strategy σ_i' which is strictly preferable to σ_i^* provided that $-i$ sticks to σ_{-i}^*. Thus a Nash equilibrium is characterized by the absence of any incentives for unilateral deviation. Of course, a Nash equilibrium can analogously be defined for a game $\langle I, S, \pi \rangle$ with only pure strategies.

---

[8] Note that in the Prisoner's Dilemma game, Kantian 'rationality' requires players to cooperate, whereas game-theoretic 'rationality' dismisses this as the irrational play of a strictly dominated strategy.

The assumption that players' rationality and the game are common knowledge does not imply that a Nash equilibrium must be played. The restriction to strategy profiles $\hat{\sigma}$ involving only strategies $\hat{\sigma}_i$ that are a best response to *some* $\sigma_{-i} \in \Delta(S_{-i})$ – also called *rationalizable strategies* – is all that can be inferred; coordination of actions and correctness of forecasts, as required by NE, is no necessary consequence. Yet, the concept has great appeal. If all players predict the same outcome of a game and this prediction is not self-destroying, then this outcome necessarily corresponds to a Nash equilibrium.[9] Also, the NE property is sufficient for a players' agreement on how to play to be self-enforcing. Moreover, the no-incentive-to-deviate property makes Nash equilibria the only candidates for stable social conventions.

Note that there need not exist a strict incentive to stick to a NE strategy σ_i^*. σ_i^* only needs to be one of possibly many *best replies* to σ_{-i}^*, i.e. $\arg\max_{\sigma_i \in \Delta(S_i)} \pi_i(\sigma_i, \sigma_{-i}^*)$ must contain σ_i^* but need not be a singleton. A Nash equilibrium σ^* where each σ_i^* constitutes the unique best reply to σ_{-i}^* is called a *strict (Nash) equilibrium*. Because a non-degenerate mixed strategy profile $\sigma \in \Delta(S_1, S_2)$ only is a Nash equilibrium if every pure strategy s_i on which σ_i places positive weight yields the same (maximal) utility given σ_{-i}, a strict equilibrium is necessarily a *pure strategy Nash equilibrium*, i.e. all players only use deterministic strategies. Pure strategy Nash equilibria do not exist for many games, but any finite game[10] has at least one mixed strategy Nash equilibrium. More generally, any game $\langle I, S, \pi \rangle$ in which all S_i are nonempty, convex, and compact subsets of a Euclidian space, and in which all π_i are continuous in s and quasi-concave in s_i, has a Nash equilibrium $s^* \in S$.

It is straightforward to extend the NE concept to extensive games of perfect information: A strategy profile σ^* is a NE of an extensive game Γ if and only if it is a NE of the normal-form representation of Γ. Both extensive games and strategic-form games often have many – sometimes implausible – Nash equilibria. Therefore, several refinements of the NE concept have been proposed.

A strategy σ_i is *completely mixed* if it places positive weight on every pure strategy $s_i \in S_i$. A strategy profile $\sigma^* \in \Delta(S_1, S_2)$ is a *(trembling hand) perfect (Nash) equilibrium (PE)* of the normal form game $\langle I, \Delta(S_1, S_2), \pi \rangle$,[11]

---

[9] Aumann and Brandenburger (1995) study sufficient epistemic conditions for a Nash equilibrium in detail. For 2-player games, *mutual knowledge* of rationality, the game, and conjectures in the form of a mixed strategy attributed to the opponent are sufficient. For more than two players, common knowledge of conjectures and a common prior are additionally required.

[10] This refers to games with a finite set of players, I, who each have a finite set of pure strategies S_i.

[11] The PE concept has originally been defined by Selten (1975) for extensive games, and establishes a refinement of both the *subgame perfect equilibrium* and *sequential equilibrium* concepts introduced below. The required *agent normal form* of an extensive game – and hence trembling-hand perfectness for extensive games – will, however, not be used in this book.

if there exists a sequence $\{\sigma^k\}_{k \in \mathbb{N}}$ of completely mixed strategy profiles converging to σ^* such that σ_i^* is a best response to σ_{-i}^k for every $k \in \mathbb{N}$. The requirement that each player's strategy is a best response to completely mixed strategies σ_{-i}^k converging to σ_{-i}^* formalizes immunity of the equilibrium against a slight perturbation in players' payoffs. Moreover, each strategy σ_i^* remains a best response also if player i assumes $-i$ to deviate from σ_{-i}^* with small probability, i. e. to 'tremble' in the implementation of the equilibrium.[12] Trivially, a NE in completely mixed strategies is perfect. A perfect equilibrium can never contain a (weakly) dominated strategy. For two-player games, any NE involving no weakly dominated strategy is perfect. A strict equilibrium is always perfect.

A second important refinement of Nash equilibria in normal-form games refers to *symmetric* two-player interaction, i. e. games $\langle I, \Delta(S', S'), \pi \rangle$ where $\pi_1(s_1, s_2) = \pi_2(s_2, s_1)$ for all $s_1, s_2 \in S'$. Such a symmetric game always has a mixed strategy *symmetric Nash equilibrium* $\sigma^* = (\sigma', \sigma')$ for $\sigma' \in \Delta(S')$, but possibly also asymmetric Nash equilibria. Consider a strategy σ' which constitutes a symmetric Nash equilibrium with itself, i. e. $(\forall \sigma'' \in S')$: $\pi_1(\sigma', \sigma') \geq \pi_1(\sigma'', \sigma')$. Then, σ' is called an *evolutionary stable strategy (ESS)* if, in addition, σ' is strictly better against any distinct σ'' than σ'' itself whenever σ'' is a best reply to σ', i. e. $\pi_1(\sigma'', \sigma') = \pi_1(\sigma', \sigma')$ implies $\pi_1(\sigma'', \sigma'') < \pi_1(\sigma', \sigma'')$ for $\sigma'' \neq \sigma'$ (Maynard Smith and Price 1973 and Maynard Smith 1974). An ESS need not exist even in finite games. Since an ESS σ' cannot involve a weakly dominated strategy, an ESS is perfect in two-player games. A strict NE is always an ESS.

Above definition of an ESS σ' is equivalent to a definition by Taylor and Jonker (1978). It requires that for any $\sigma'' \neq \sigma'$ there exists some $\varepsilon_{\sigma''} \in (0, 1)$ such that for all $\varepsilon \in (0, \varepsilon_{\sigma''})$ the strict inequality $\pi_1(\sigma', \varepsilon \sigma'' + (1 - \varepsilon)\sigma') > \pi_1(\sigma'', \varepsilon \sigma'' + (1 - \varepsilon)\sigma')$ holds.[13] Intuitively, if an ESS σ' is played by all members of a monomorphic population of biological or social agents and a group of 'mutants' playing σ'' enters with less than $\varepsilon_{\sigma''}$ share of the resulting population, their hypothetical 'invasion' would be repelled if payoffs correspond to reproductive success. A *neutrally stable strategy (NSS)* σ' needs not actively repel invasions, but cannot be overrun by isolated mutants. Formally, a NSS is defined by satisfying the above inequality weakly.

The main Nash equilibrium refinement for extensive games deals with the credibility problem of NE strategies σ_i^* which are sub-optimal if, for whatever reasons, i should be called to move in an information set un-reached by σ^*. A *subgame* of an extensive game Γ with decision nodes $\mathcal{D}$, information sets $\mathcal{H}$, and precedence relation $\prec$ is an extensive game Γ_n with the following properties. First, it has a decision node $n \in \mathcal{D}$ with $\{n\} \in \mathcal{H}$ as its initial node.

---

[12] Various further refinements based on other specifications of 'trembles' have been proposed. Van Damme (1991) is the most comprehensive reference also on refinements of other equilibrium concepts.

[13] In fact, a uniform *invasion barrier* $\bar{\varepsilon} = \varepsilon_{\sigma''}$ can be used for all $\sigma'' \neq \sigma'$ in finite symmetric games.

Second, it has exactly the successors of n, $S(n)$, as additional nodes. Third, every information set $H \in \mathcal{H}$ with $n' \in H$ for $n' \in S(n)$ is a subset of $S(n)$.[14] Finally, everything else – including precedence relation, information partition, payoffs, etc. – is inherited from Γ. A strategy profile σ^* for the extensive game Γ is a *subgame perfect (Nash) equilibrium (SPE)* if it induces a Nash equilibrium in every subgame of Γ (Selten 1965). Since Γ itself is a subgame of Γ, SPE implies NE. SPE adds *sequential rationality* to the rational behaviour formalized by NE. The underlying assumption that each player continues to believe his opponent to be rational even after the observation of an irrational move is not uncontroversial.

Harsanyi's (1967/68) extension of the NE concept to Bayesian games $\langle I, S, \Theta, F_\Theta, \pi \rangle$ requires from a strategy profile s^* that each $s_i^*(\theta_i)$ is for every player $i \in I$ and type $\theta_i \in \Theta_i$ a best response to s_{-i}^* when expectations over θ_{-i} are computed conditional on i's realized θ_i using Bayes' rule. Formally, a pure strategy profile[15] s^* is a *Bayesian (Nash) equilibrium*, if $(\forall i \in I)(\forall \theta_i \in \Theta_i)(\forall s_i' \in S_i)$: $E_{\theta_{-i}} [\pi_i(s_i^*(\theta_i), s_{-i}(\theta_{-i}), \theta_i) | \theta_i] \geq E_{\theta_{-i}} [\pi_i(s_i', s_{-i}(\theta_{-i}), \theta_i) | \theta_i]$. A mixed strategy Nash equilibrium σ^* of a game $\langle I, \Delta(S_1, S_2), \pi \rangle$ of complete information can be interpreted as the pure strategy Bayesian equilibrium of a Bayesian game with a continuum of types for each player i, which all have identical preferences but choose θ_i-dependent pure strategies $s_i(\theta_i)$, with population shares corresponding to the weight placed on $s_i(\theta_i)$ by σ_i^*.

In extensive games of imperfect information – also called *Bayesian extensive games* – subgame perfectness is no longer guaranteed to rule out incredible threats or promises.[16] It is in the spirit of subgame perfectness to require that each player acts optimally in every information set given subjective, but rational beliefs about how this set was reached. Player i's *beliefs* $\mu(H)$ for information set $H \in \mathcal{H}_i$ are formally a probability distribution over the nodes $n \in H$. An equilibrium is in this context most conveniently specified as the combination of a behavioural strategy profile $b = (b_1, b_2) \in B = B_1 \times B_2$, and a *belief system* $\mu = \{\mu(H)\}_{H \in \mathcal{H}}$. A pair (b, μ) is called an *assessment*.

For information sets which have positive probability of being reached under a strategy profile b, beliefs are rationally formed using *Bayes' rule*, which defines the probability of event A_1 conditional on the positive-probability event A_2 as $\text{Prob}(A_1 | A_2) = \text{Prob}(A_1 \wedge A_2) / \text{Prob}(A_2)$. Players, however, have great freedom in information sets that have zero probability of being reached, since conditional probabilities cannot be deduced from b. Various attempts have been made to impose plausible restrictions on beliefs after zero-probability events. Clearly, no problem arises in case that b is a *completely*

---

[14] Therefore, in games of imperfect information, a subgame Γ_n cannot be constructed for all nodes $n \in \mathcal{D}$.

[15] A Bayesian equilibrium in mixed strategies can be defined analogously.

[16] This is because optimality of a strategy is only required in sub-trees starting at a singleton information set $\{n\}$. Often this means that the game itself is its only subgame.

mixed behavioural strategy profile, i. e. every action at every information set has a positive probability of being played. Kreps and Wilson (1982) consider an assessment (b, μ) to have particular plausibility if it is in some sense supported by nearby completely mixed strategy profiles. Formally, they define an assessment (b, μ) to be *consistent* if there exists a sequence $\{(b^k, \mu^k)\}_{k \in \mathbb{N}}$ of assessments converging to (b, μ) where each b^k is completely mixed and μ^k is derived from b^k by Bayesian updating. Let $E[\pi_i \mid H, \mu(H), b]$ denote player i's expected payoff conditional on information set H being reached with the probability distribution $\mu(H)$ over nodes $n \in H$ and future moves being specified by behavioural strategy profile b. An assessment (b, μ) is *sequentially rational* if $(\forall i \in I)\, (\forall H \in \mathcal{H}_i)\, (\forall b_i' \in B_i)\colon E[\pi_i \mid H, \mu(H), b_i, b_{-i}] \geq E[\pi_i \mid H, \mu(H), b_i', b_{-i}]$, which formalizes optimal behaviour in every information set given belief system μ. An assessment (b^*, μ^*) is a *sequential equilibrium (SE)* if it is consistent and sequentially rational.

Sequential equilibrium and subgame perfect equilibrium coincide for extensive games of perfect information, where the consistent belief system is trivially $\mu(H)(n) = 1$ for $H = \{n\}$, and $\mu(H)(n) = 0$ otherwise, for all $H \in \mathcal{H}$. Every finite extensive game with perfect recall has a SE, and hence a SPE.

A *Bayesian extensive game with observable actions* is an extensive game in which the only unobserved move is Nature's initial draw of independent player types according to a product measure ϱ on Θ. An assessment (b^*, μ^*) is called a *perfect Bayesian equilibrium (PBE)* if, first, it is sequentially rational. Second, beliefs are based on ϱ and Bayesian updating wherever that is possible.[17] And, third, $-i$'s beliefs about θ_i are only affected by $-i$'s actions. In finite Bayesian extensive games with observable actions, a SE (b^*, μ^*) is also a PBE. The reverse is true if there are at most two possible types for each player, or if each player only moves once. PBE is often easier to work with than SE.

The Nash equilibrium concept and its various refinements describe, or even prescribe, behaviour of rational players who consciously maximize a possibly complicated utility function given beliefs about opponents' behaviour which – granted that everybody acts according to the equilibrium strategy profile – turn out to be correct. It is a natural question whether the quite demanding NE concept is relevant also under weaker behavioural assumptions? It can be answered in the affirmative.

Assume a single continuous population of agents that are each programmed to play a particular pure strategy in a finite symmetric normal form game $\langle I, \Delta(S'), S'), \pi \rangle$ with each other. A mixed strategy $\sigma' \in \Delta(S')$ corresponds to a *population state* $x(t) \in \Delta(S')$ at time t, where the weight placed by σ' on pure strategy $s^j \in S'$ is the population share,

---

[17] This means up to an information set where some player i's action contradicts his strategy b_i^*, and again, after a new arbitrary distribution on Θ_i has been assumed by $-i$ (and from there on until the next move which contradicts b^*, etc.).

$x_j(t)$, of agents playing s^j. The payoffs to agents using pure strategy s^j against randomly drawn individuals, or equivalently the entire population, is $\pi_1(s^j, x(t))$. It is interpreted as the incremental effect[18] on the number of offspring, called *fitness*. Assuming that reproduction takes place continuously over time, this defines the *(single population) replicator dynamics* $\dot{x}_j(t) = x_j(t) \left[\pi_1(s^j, x(t)) - \pi_1(x(t), x(t)) \right]$ for $j = 1, \ldots, |S|$ (Taylor and Jonker 1978).

A *rest point* of the replicator dynamics is a population state x^* with $\dot{x}(t) = 0$ for $x(t) = x^*$, and is also called a *stationary state* or *dynamic equilibrium*. A degenerate state x in which the entire population plays a unique pure strategy is necessarily a rest point. Also, all strategies σ' that form a symmetric NE $\sigma^* = (\sigma', \sigma')$ are rest points. Any non-degenerate – meaning interior – rest point of the replicator dynamics forms a NE.

A dynamic equilibrium $x^* \in \Delta(S')$ is *Lyapunov stable* if dynamics from states x' nearby do not lead far away from x^*, or, formally, if every neighbourhood U_{x^*} of x^* contains a neighbourhood $U'_{x^*} \subseteq U_{x^*}$ such that $(\forall x^0 \in U'_{x^*}): \{x(0) = x^0 \implies x(t) \in U_{x^*}\}$ for $t > 0$. The rest point x^* is *asymptotically stable* if it is Lyapunov stable and there exists a neighbourhood $U^*_{x^*}$ such that $(\forall x^0 \in U^*_{x^*}): \{x(0) = x^0 \implies \lim_{t \to \infty} x(t) = x^*\}$, i.e. any *trajectory* of population states, $\{x(t)\}_{t \geq 0}$, which starts close to x^* stays close to x^* and eventually converges to it. This means that an asymptotically stable population state is immune to isolated small perturbations, e.g. the introduction of a small number of mutants.

Any Lyapunov stable dynamic equilibrium forms a NE. Any asymptotically stable dynamic equilibrium forms a PE. Any ESS is asymptotically stable, and any NSS is Lyapunov stable. The reverses are not generally true. If the initial population state is completely mixed, any pure strategy $s' \in S'$ that can be deleted from S' by the iterative elimination of strictly dominated strategies vanishes for $t \to \infty$.[19] A strategy s^j that is only weakly dominated by some s^k may survive, provided that all pure strategies against which s^k is strictly better than s^j vanish. If the initial population state is completely mixed and $x(t)$ does converge to some state x^* for $t \to \infty$, then x^* is a Nash equilibrium.[20]

Other evolutionary dynamics – involving less specific selection equations, multiple populations of agents, and the consideration also of asymmetric games – more or less confirm these findings. The replicator equation and its

---

[18] This assumes that all individuals have the same baseline birthrate, and also an identical death rate.

[19] This equivalence between the result of rational reasoning and the long-run implications of evolution need not be valid if replicator dynamics are modelled in discrete time.

[20] $x(t)$ need not converge to some x, but may, for example, enter a limit cycle. Convergence to x^* for some $x(0) \neq x^*$ implies that x^* is stationary. Even if convergence occurs for every interior initial point, this does, however, not imply Lyapunov stability of x^*.

generalizations arise also in non-biological models of imitation or aspiration learning (see Sect. 1.4). Thus there is often an evolutionary foundation of Nash equilibrium and its refinements complementing the rationalistic reasons for their application.

List of Symbols

$\mathbb{R}$	set of real numbers
$\mathbb{R}_+$, $\mathbb{R}_{++}$	set of non-negative (and positive) real numbers
$\mathbb{N}$	set of natural numbers, i.e. $\{1, 2, 3, \ldots\}$
$\wp(I)$	set of subsets of I
$x \geqq y$	$(\forall i \in I)\colon x_i \geq y_i$
$x \geq y$	$x \geqq y$ and $(\exists i \in I)\colon x_i > y_i$
$x > y$	$(\forall i \in I)\colon x_i > y_i$
$\lfloor x \rfloor$	biggest integer smaller than or equal to x
$\lceil x \rceil$	smallest integer bigger than or equal to x
$x \cdot y$	inner product of vectors x and y
∇f	column vector of partial derivatives of function f
$^\lambda \text{Prob}(A)$	λ-dependent probability of event A
$p_i \overset{i.i.d.}{\sim} U[0,1]$	p_i is drawn from $U[0,1]$ independently of p_j $(j \neq i)$
$A \setminus B$	difference between set A and set B
$|A|$	cardinality of set A
a, b	personal utils of player 1 (and 2) worth one social util
c	a constant
c_i	cost of one stage of bargaining to player i
c_x	memory of the m-fold repetition of $(x, 1-x)$
$c_{s_1 s_2}$	a state in which $l_i = \pi_i(s_1, s_2)$ for $i = 1, 2$
f_i	partial derivative of f with respect to p_i
g_i	cond. density of player i's post-perturbation aspiration
h_t	society's memory in period t, a state of Φ^0 and Φ^η
i	a player
$-i$	collection of players $j \neq i$ in I
l_i, l_i^t	aspiration level of player i (in period t)
l_i^*	aspiration update function of player i
m	length of society's memory / cardinality of $I^*(v)$
n	number of bargaining stages / cardinality of I
o	a deterministic outcome
p_i	inertia function of player i / acceptance rate of player i

$\tilde{p}_i$	lower bound on player i's inertia
$p_{h\,h'}^{(t)}$	probability to move from state h to h' in t steps
$q,\ q_0$	state of an automaton, the initial state
r_i	sampling ratio of player i
r_{kl}	resistance of a transition from recurrence class C_k to C_l
s	a pure strategy profile
s_i	a pure strategy of player i
t	time or period index
$u_i(x)$	immediate utility of deterministic agreement x to player i
$u_i^B(U)$	player i's maximal payoff in U
u^D	disagreement point or status quo point
v	(characteristic function of) a simple game
$v_i(x,t)$	present value of agreement x at time t for player i
x_t	proposal for player 1's share / system state in period t
$x(t)$	vector of population shares / a player's demand at time t
$x_j(t)$	share of the population playing strategy s^j at time t
B	exogenous breakdown of negotiations
C	set of conventions $c_{s_1 s_2}$ of the satisficing process
C_k	a recurrence class of Φ^0
$C_i(v)$	set of coalitions in which player i is crucial
D	outcome resulting from final or perpetual disagreement
E	state space of satisficing processes / expectation operator
F_{θ_i}	cumulative distribution function of i's type
$F^E,\ F^{E(a,b)}$	egalitarian and weighted egalitarian bargaining solution
F^{KS}	Kalai-Smorodinsky bargaining solution
F^N	Nash bargaining solution (NBS)
$F^{N(\alpha,\beta)}$	asymmetric NBS with bargaining powers α and β
F^R	Rawlsian bargaining solution
$F^U,\ F^{U(b,a)}$	utilitarian and weighted utilitarian bargaining solution
H	state space of adaptive play processes / a high offer
I	set of players
$I^*(v)$	set of inferior players in simple game v
I_i	population of agents acting in the role of player i
L	a low offer
L_i	set of player i's aspiration levels
M	a machine or finite state automaton
$M_{(1)},\ M_{(2)}$	automata acting in the role of player 1 (and player 2)
M_i	slope parameter of inertia function p_i
$M(v)$	set of minimal winning coalitions of simple game v
N	rejection of a low offer

O	set of deterministic outcomes
O_i	outcome resulting from player i choosing to opt out
P	transition probability matrix / transition kernel of Φ^0
P^η	transition probability kernel of Φ^η
$P(U)$	set of Pareto-efficient elements of U
Q_M	set of states of automaton M
Q, Q_i, Q_*	conditional transition probability kernels of Φ^η
R	'fast forward' transition kernel of Φ^0, i.e. $\lim_{n\to\infty} P^n$
S	set of pure strategy profiles / a coalition
S_i	set of pure strategies of player i
T	set of agreement times / a distinguished period
U	set of feasible personal payoff combinations
$\tilde{U}$	set of feasible empathetic payoff combinations
$U[0,1]$	uniform distribution on $[0,1]$
$W(v)$	set of winning coalitions of simple game v
X	set of proposals, usually $[0,1]$
X^t	set of bargaining histories involving t proposals
X_ξ	discrete set of proposals which are multiples of ξ
Y	acceptance of a low offer / a Borel subset of E
$\mathcal{G}^I$	set of simple games with the set of players I
$\mathcal{I}$	set of inputs of an automaton
$\mathcal{M}$	set of finite bargaining automata
$\mathcal{N}$	a component of mixed-strategy Nash equilibria
$\mathcal{O}$	set of outputs of an automaton
$\mathcal{R}$	set of Rubinstein bargaining games (RBG)
$\mathcal{R}_i$	set of subgames of a RBG starting with player i's offer
α	bargaining power of player 1
β	bargaining power of player 2 / Banzhaf power index
δ_i	discount factor of player i
η, η_i	probability of a perturbation (in population i)
$\eta_i(v)$	number of swings of player i in simple game v
$\tilde{\eta}_i(v)$	number of strict swings of player i in simple game v
$\eta_i^{(\theta)}(v)$	number of θ-swings of player i in simple game v
θ_i	player i's type
λ_i	noise scaling factor / aspiration persistence of player i
μ	a power index
μ_i	collection of i's beliefs
μ_t	distribution over state space H in period t
μ^η	stationary distribution (invariant measure) of Φ^η
μ^*	limit of μ^η for $\eta \to 0$

ξ	(possible) smallest unit in which offers have to be made
π_i	von Neumann-Morgenstern personal utility of player i
ϱ	a permutation function on I
σ	a mixed strategy profile / the strict power index (SPI)
σ_i	a mixed strategy of player i / standard deviation
$\sigma(E)$	Borel σ-algebra of E
σ^c	the generalized strict power index (GSPI)
ς_M	transition function of automaton M
τ	length of one bargaining stage
τ_i	strictly increasing affine transformation of agent i's utility
υ_i	von Neumann-Morgenstern empathetic utility of player i
φ	the Shapley-Shubik index of power (SSI)
χ_M	output function of automaton M
Γ	a game / the game of life
Γ^∞	the repeated game of life
$\bar{\Gamma}^\infty$	the repeated game of morals
Δ_i^t	disappointment of player i in period t
$\Delta(O)$	set of probability distributions over O
$\Delta(S_1, S_2)$	set of mixed strategy profiles
$\Delta(S_i)$	set of mixed strategies of player i
Θ_i	set of player i's types
Υ_i	player i's expected empathetic utility
$\mathfrak{S}$	a random coalition, i.e. a collection of acceptance rates
$\succcurlyeq$	complexity ordering on $\mathcal{M}$
$\succsim_i$	personal preference ordering of player i
$\succsim^i$	empathetic preference ordering of player i
$>_i$	preference ordering of player i on $\mathcal{G}^I$

List of Figures

References

Aumann, R. and A. Brandenburger (1995). Epistemic conditions for Nash equilibrium. *Econometrica 63*(5), 1161–1180.

Aumann, R. and M. Maschler (1964). The bargaining set for cooperative games. In L. S. Shapley, M. Dresher, and A. Tucker (Eds.), *Advances in Game Theory*, pp. 443–476. Princeton, NJ: Princeton University Press.

Aumann, R. J. and S. Hart (Eds.) (1994). *Handbook of Game Theory*, Volume II. Amsterdam: Elsevier Science.

Baldwin, R. E., E. Berglöf, F. Giavazzi, and M. Widgrén (2000). EU reforms for tomorrow's Europe. CEPR Discussion Paper 2623, Centre for Economic Policy Research.

Baldwin, R. E., J. F. Francois, and R. Portes (1997). The costs and benefits of eastern enlargement: The impact on the EU and central Europe. *Economic Policy 24*, 125–176.

Banzhaf, J. F. (1965). Weighted voting doesn't work: A mathematical analysis. *Rutgers Law Review 19*(2), 317–343.

Barlow, R. E. and F. Proschan (1975). *Statistical Theory of Reliability and Life Testing: Probability Models*. New York: Holt, Rinehart and Winston.

Baron, D. P. and J. A. Ferejohn (1989). Bargaining in legislatures. *American Political Science Review 83*(4), 1181–1206.

Benaim, M. and J. W. Weibull (2000). Deterministic approximation of stochastic evolution in games. Working Paper 534, Research Institute of Industrial Economics, Stockholm.

Bentham, J. (1789). *An Introduction to the Principles of Morals and Legislation*. In J. Bowring (Ed.), *The Works of Jeremy Bentham*, Volume 1. London, 1838-1843. Reprinted by Russell & Russell, New York, 1962.

Bergin, L. and B. L. Lipman (1996). Evolution with state-dependent mutations. *Econometrica 64*(4), 943–956.

Berninghaus, S. K. and U. Schwalbe (1996). Conventions, local interaction, and automata networks. *Journal of Evolutionary Economics 6*(3), 297–312.

Binmore, K. G. (1987a). Nash bargaining theory II. See Binmore and Dasgupta (1987a), pp. 61–76.

Binmore, K. G. (1987b). Perfect equilibria in bargaining models. See Binmore and Dasgupta (1987a), pp. 77–105.

Binmore, K. G. (1992). *Fun and Games.* Lexington, MA: D.C. Heath.

Binmore, K. G. (1994). *Game Theory and the Social Contract – Volume I: Playing Fair.* Cambridge, MA: MIT Press.

Binmore, K. G. (1998a). The evolution of fairness norms. *Rationality and Society 10*(3), 275–301.

Binmore, K. G. (1998b). *Game Theory and the Social Contract – Volume II: Just Playing.* Cambridge, MA: MIT Press.

Binmore, K. G. (2001). Evolutionary social theory: Reply to Robert Sugden. *Economic Journal 111*(469), F244–248.

Binmore, K. G. and P. Dasgupta (Eds.) (1987a). *The Economics of Bargaining.* Oxford: Basil Blackwell.

Binmore, K. G. and P. Dasgupta (1987b). Nash bargaining theory: An introduction. See Binmore and Dasgupta (1987a), pp. 1–26.

Binmore, K. G., J. McCarthy, G. Ponti, L. Samuelson, and A. Shaked (1999). A backward induction experiment. SSRI Working Paper 9934, University of Wisconsin, Madison. Forthcoming in *Journal of Economic Theory.*

Binmore, K. G., P. Morgan, A. Shaked, and J. Sutton (1991). Do people exploit their bargaining power? An experimental study. *Games and Economic Behavior 3*(3), 295–322.

Binmore, K. G., M. J. Osborne, and A. Rubinstein (1992). Noncooperative models of bargaining. In R. J. Aumann and S. Hart (Eds.), *Handbook of Game Theory,* Volume I, pp. 179–225. Amsterdam: North-Holland.

Binmore, K. G., M. Piccione, and L. Samuelson (1998). Evolutionary stability in alternating-offers bargaining games. *Journal of Economic Theory 80*(2), 257–291.

Binmore, K. G., A. Rubinstein, and A. Wolinsky (1986). The Nash bargaining solution in economic modelling. *RAND Journal of Economics 17*(2), 176–188.

Binmore, K. G., A. Shaked, and J. Sutton (1985). Testing noncooperative bargaining theory: A preliminary study. *American Economic Review 75*(5), 1178–1180.

Binmore, K. G., A. Shaked, and J. Sutton (1989). An outside option experiment. *Quarterly Journal of Economics 104*(4), 753–770.

Binmore, K. G., J. Swierzbinski, S. Hsu, and C. Proulx (1993). Focal bargaining. *International Journal of Game Theory 22*(4), 381–409.

Birnbaum, Z. W. (1969). On the importance of different components in a multicomponent system. In P. R. Krishnaiah (Ed.), *Multivariate Analysis – II,* pp. 581–592. New York: Academic Press.

Bishop, R. L. (1963). Game-theoretic analyses of bargaining. *Quarterly Journal of Economics 77*(4), 559–602.

Börgers, T. and R. Sarin (2000). Naive reinforcement learning with endogenous aspirations. *International Economic Review 41*(4), 921–950.

Bowley, A. L. (1924). *The Mathematical Groundwork of Economics.* Reprinted by A. M. Kelley, New York, 1967.

Brams, S. J. and A. D. Taylor (1996). *Fair Division - From Cake-Cutting to Dispute Resolution*. Cambridge: Cambridge University Press.

Camerer, C. and T.-H. Ho (1999). Experience-weighted attraction learning in normal form games. *Econometrica 67*(4), 827–874.

Camerer, C., E. J. Johnson, T. Rymon, and S. Sen (1993). Cognition and framing in sequential bargaining for gains and losses. In K. G. Binmore, A. Kirman, and P. Tani (Eds.), *Frontiers of Game Theory*, pp. 27–47. Cambridge, MA: MIT Press.

Camerer, C. and R. H. Thaler (1995). Anomalies - Ultimatums, dictators and manners. *Journal of Economic Perspectives 9*(2), 209–219.

Cameron, L. A. (1999). Raising the stakes in the ultimatum game: Experimental evidence from Indonesia. *Economic Inquiry 37*(1), 47–59.

Chatterjee, K., B. Dutta, D. Ray, and K. Sengupta (1993). A noncooperative theory of coalitional bargaining. *Review of Economic Studies 60*(2), 463–477.

Chatterjee, K. and H. Sabourian (2000). Multiperson bargaining and strategic complexity. *Econometrica 68*(6), 1491–1509.

Cross, J. G. (1965). A theory of the bargaining process. *American Economic Review 55*(1), 67–94.

Deegan, J. and E. W. Packel (1978). A new index of power for simple n-person games. *International Journal of Game Theory 7*(2), 113–123.

Dixon, H. (2000). Keeping up with the Joneses: Competition and the evolution of collusion. *Journal of Economic Behavior and Organization 43*(2), 223–238.

Dodds, W. (1973). An application of the Bass model in long-term new product forecasting. *Journal of Marketing Research 10*(3), 308–311.

Dubey, P. and L. Shapley (1979). Mathematical properties of the Banzhaf power index. *Mathematics of Operations Research 4*(2), 99–131.

Eatwell, J., M. Milgate, and P. Newman (Eds.) (1991). *The New Palgrave - The World of Economics*. London: Macmillan. First published in *The New Palgrave: A Dictionary of Economics*, London: Macmillan, 1987.

Edgeworth, F. Y. (1881). *Mathematical Psychics: An Essay on the Application of Mathematics to the Moral Sciences*. Reprinted by A. M. Kelley, New York, 1961.

Ellingsen, T. (1997). The evolution of bargaining behaviour. *Quarterly Journal of Economics 112*(2), 581–602.

Ellison, G. (1993). Learning, local interaction, and coordination. *Econometrica 61*(5), 1047–1071.

Felsenthal, D. and M. Machover (1998). *The Measurement of Voting Power - Theory and Practice, Problems and Paradoxes*. Cheltenham: Edward Elgar.

Fernandez, R. and J. Glazer (1991). Striking for a bargain between two completely informed agents. *American Economic Review 81*(1), 240–252.

Fishburn, P. C. and A. Rubinstein (1982). Time preference. *International Economic Review* 23(3), 677–694.

Foster, D. and H. P. Young (1990). Stochastic evolutionary game dynamics. *Theoretical Population Biology* 38(2), 219–232.

Freidlin, M. and A. Wentzell (1984). *Random Perturbations of Dynamical Systems*. New York: Springer-Verlag.

Fudenberg, D. and D. K. Levine (1998). *The Theory of Learning in Games*. Cambridge, MA: MIT Press.

Fudenberg, D. and J. Tirole (1991). *Game Theory*. Cambridge, MA: MIT Press.

Gale, J., K. G. Binmore, and L. Samuelson (1995). Learning to be imperfect: The ultimatum game. *Games and Economic Behavior* 8(1), 56–90.

Gambarelli, G. and G. Owen (1994). Indirect control of corporations. *International Journal of Game Theory* 23(4), 287–302.

Garret, G. and G. Tsebelis (1999). Why resist the temptation to apply power indices to the European Union. *Journal of Theoretical Politics* 11(3), 291–308.

Gilboa, I. and D. Schmeidler (1996). Case-based optimization. *Games and Economic Behavior* 15(1), 1–26.

Gul, F. (1989). Bargaining foundations of Shapley value. *Econometrica* 57(1), 81–95.

Güth, W. (1995). On ultimatum bargaining experiments – A personal review. *Journal of Economic Behavior and Organization* 27(3), 329–344.

Güth, W., S. Huck, and P. Ockenfels (1996). Two-level ultimatum bargaining with incomplete information: An experimental study. *Economic Journal* 106(436), 593–604.

Güth, W., R. Schmittberger, and B. Schwarze (1982). An experimental analysis of ultimatum bargaining. *Journal of Economic Behavior and Organization* 3(4), 367–388.

Güth, W. and E. van Damme (1998). Information, strategic behavior, and fairness in ultimatum bargaining: An experimental study. *Journal of Mathematical Psychology* 42(2-3), 227–247.

Gutsche, J. (2000). Ökonomische Theorie der Verhandlungen. Diplomarbeit, WIOR, Universität Karlsruhe.

Haller, H. and S. Holden (1990). A letter to the editor on wage bargaining. *Journal of Economic Theory* 52(1), 232–236.

Harms, W. (1997). Evolution and ultimatum bargaining. *Theory and Decision* 42(2), 147–176.

Harrison, G. W. and K. McCabe (1996). Expectations and fairness in a simple bargaining experiment. *International Journal of Game Theory* 25(3), 303–327.

Harsanyi, J. C. (1953). Cardinal utility in welfare economics and in the theory of risk-taking. *Journal of Political Economy* 61(5), 434–435.

Harsanyi, J. C. (1956). Approaches to the bargaining problem before and after the theory of games: A critical discussion of Zeuthen's, Hicks', and Nash's theories. *Econometrica 24*(2), 144–157.

Harsanyi, J. C. (1975). Can the maximin principle serve as a basis for morality? A critique of John Rawls' theory. *American Political Science Review 69*, 594–606.

Harsanyi, J. C. (1977). *Rational Behavior and Bargainig Equilibrium in Games and Social Situations.* Cambridge: Cambridge University Press.

Harsanyi, J. C. (1987). Interpersonal utility comparisons. See Eatwell, Milgate, and Newman (1991), pp. 361–366. First published in *The New Palgrave: A Dictionary of Economics*, London: Macmillan, 1987.

Harsanyi, J. C. (1967/68). Games with incomplete information played by 'Bayesian' players, Parts I, II, and III. *Management Science – Theory 14*, 159–182, 320–334, and 486–502.

Harsanyi, J. C. and R. Selten (1972). A generalized Nash solution for two-person bargaining games with incomplete information. *Management Science – Theory 18*(5), P80–P106.

Hart, S. and A. Mas-Colell (1996). Bargaining and value. *Econometrica 64*(2), 357–380.

Hehenkamp, B. (1999). Sluggish consumers: An evolutionary solution to the Bertrand paradox. Discussion Paper 99-04, Wirtschafts- und Sozialwissenschaftliche Fakultät, Universität Dortmund. Forthcoming in *Games and Economic Behavior*.

Hicks, J. R. (1932). *The Theory of Wages.* New York: Macmillan.

Holler, M. J. (1978). A priori party power and government formation. *Munich Social Science Review 4*, 25–41.

Holler, M. J. (1982a). Party power and government formation: A case study. See Holler (1982b), pp. 273–282.

Holler, M. J. (Ed.) (1982b). *Power, Voting, and Voting Power.* Würzburg: Physica-Verlag.

Holler, M. J. (1992). *Ökonomische Theorie der Verhandlungen* (3rd ed.). München: Oldenbourg-Verlag.

Holler, M. J. (1996). Review of "Binmore, K. G. (1994). Game Theory and the Social Contract – Volume I: Playing Fair. Cambridge, MA: MIT Press.". *Journal of Economics (Zeitschrift für Nationalökonomie) 63*, 102–105.

Holler, M. J. (1997). Power, monotonicity and expectations. *Control and Cybernetics 26*, 605–607.

Holler, M. J. (2000). Review of "Binmore, K. G. (1998). Game Theory and the Social Contract – Volume II: Just Playing. Cambridge, MA: MIT Press.". *Journal of Economics (Zeitschrift für Nationalökonomie) 71*(2), 200–204.

Holler, M. J. and G. Illing (2000). *Einführung in die Spieltheorie* (4th ed.). Berlin: Springer-Verlag.

Holler, M. J. and S. Napel (2001). On interpersonal comparison of value. In K. Nevalainen (Ed.), *Justice, Charity, and the Welfare State: Moral and Social Dimensions*, Acta Philosophica Fennica *68*, pp. 119–138. Helsinki: Societas Philosophica Fennica.

Holler, M. J. and G. Owen (Eds.) (2001). *Power Indices and Coalition Formation*. Dordrecht: Kluwer.

Holler, M. J. and E. W. Packel (1983). Power, luck and the right index. *Zeitschrift für Nationalökonomie (Journal of Economics) 43*(1), 21–29.

Holler, M. J. and M. Widgrén (1999). Why power indices for assessing EU decision-making. *Journal of Theoretical Politics 11*(3), 321–330.

Huck, S. and J. Oechssler (1999). The indirect evolutionary approach to explaining fair allocations. *Games and Economic Behavior 28*(1), 13–24.

Josephson, J. and A. Matros (2000). Stochastic imitation in finite games. Working Paper 363, Stockholm School of Economics.

Kaarbøe, O. M. and A. F. Tieman (1999). Equilibrium selection under different learning modes in supermodular games. Discussion Paper 1299, Department of Economics, University of Bergen. (Revised: Evolution with mutations driven by experimentation; December 2000).

Kagel, J. H. and A. E. Roth (Eds.) (1995). *Handbook of Experimental Economics*. Princeton, NJ: Princeton University Press.

Kahnemann, D. and A. Tversky (1979). Prospect theory: An analysis of decision under risk. *Econometrica 47*(2), 263–291.

Kalai, E. (1977a). Nonsymmetric Nash solutions and replications of 2-person bargaining. *International Journal of Game Theory 6*(3), 129–133.

Kalai, E. (1977b). Proportional solutions to bargaining situations: Interpersonal utility comparisons. *Econometrica 45*(7), 1623–1630.

Kalai, E. and D. Samet (1987). On weighted Shapley values. *International Journal of Game Theory 16*(3), 205–222.

Kalai, E. and M. Smorodinsky (1975). Other solutions to Nash's bargaining problem. *Econometrica 43*(3), 513–518.

Kandori, M., G. J. Mailath, and R. Rob (1993). Learning, mutation, and long run equilibria in games. *Econometrica 61*(1), 29–56.

Karandikar, R., D. Mookherjee, D. Ray, and F. Vega-Redondo (1998). Evolving aspirations and cooperation. *Journal of Economic Theory 80*(2), 292–331.

Kennan, J. and R. Wilson (1993). Bargaining with private information. *Journal of Economic Literature 31*(1), 45–104.

Kihlstrom, R. E., A. E. Roth, and D. Schmeidler (1981). Risk aversion and solutions to Nash's bargaining problem. In O. Moeschlin and D. Pallaschke (Eds.), *Game Theory and Mathematical Economics*, pp. 65–71. Amsterdam: North-Holland.

Kim, Y. (1999). Satisficing and optimality in 2×2 common interest games. *Economic Theory 13*(2), 365–375.

Kirman, A. and M. Widgrén (1995). European economic decision-making policy: Progress or paralysis? *Economic Policy 21*, 421–460.

Krelle, W. (1976). *Preistheorie*, Volume II. Tübingen: J. C. B. Mohr (Paul Siebeck).

Kreps, D. M. and R. Wilson (1982). Sequential equilibria. *Econometrica 50*(4), 863–895.

Krishna, V. and R. Serrano (1995). Perfect equilibria of a model of *n*-person noncooperative bargaining. *International Journal of Game Theory 24*(3), 259–272.

Kuhn, H. W. (1953). Extensive games and the problem of information. In H. W. Kuhn and A. W. Tucker (Eds.), *Contributions to the Theory of Games*, Volume II, pp. 193–216. Princeton, NJ: Princeton University Press.

Kultti, K. (1994). An *n*-person bargaining game. *Finnish Economic Papers 7*(2), 130–132.

Lant, T. K. (1992). Aspiration level adaptation: An empirical exploration. *Management Science 38*(5), 623–644.

Laruelle, A. and F. Valenciano (2001). Shapley-Shubik and Banzhaf indices revisited. *Mathematics of Operations Research 26*(1), 89–104.

Laruelle, A. and M. Widgrén (1998). Is the allocation of power among EU states fair? *Public Choice 94*(3-4), 317–340.

Leech, D. (1988). The relationship between shareholding concentration and shareholder voting power in British companies: A study of the application of power indices. *Management Science 34*(4), 509–527.

Levínský, R. and P. Silárszky (2001). Global monotonicity of values of cooperative games. See Holler and Owen (2001), pp. 105–125.

Linster, B. G. (2000). Review of "Ken Binmore, Just Playing. Game Theory and the Social Contract, vol. 2. Cambridge: MIT Press, 1998.". *Public Choice 102*(1), 172–175.

Lopes, L. L. (1996). When time is of the essence: Averaging, aspiration, and the short run. *Organizational Behavior and Human Decision Processes 65*(3), 179–189.

Luce, R. D. and H. Raiffa (1957). *Games and Decisions*. New York: Wiley.

Mas-Colell, A., M. D. Whinston, and J. R. Green (1995). *Microeconomic Theory*. Oxford: Oxford University Press.

Maschler, M. and B. Peleg (1966). A characterization, existence proof and dimension bounds for the kernel of a game. *Pacific Journal of Mathematics 18*(2), 289–328.

Maynard Smith, J. (1974). The theory of games and the evolution of animal conflicts. *Journal of Theoretical Biology 47*, 209–221.

Maynard Smith, J. and G. R. Price (1973, November 2). The logic of animal conflict. *Nature 246*, 15–18.

Meyn, S. P. and R. L. Tweedie (1993). *Markov Chains and Stochastic Stability*. London: Springer-Verlag.

Mookherjee, D. and B. Sopher (1994). Learninig behavior in an experimental matching pennies game. *Games and Economic Behavior* 7(1), 62–91.

Mookherjee, D. and B. Sopher (1997). Learning and decision costs in experimental constant sum games. *Games and Economic Behavior* 19(1), 97–132.

Murnighan, J. K., A. E. Roth, and F. Schoumaker (1988). Risk aversion in bargaining: An experimental study. *Journal of Risk and Uncertainty* 1(1), 101–124.

Muthoo, A. (1999). *Bargaining Theory with Applications*. Cambridge: Cambridge University Press.

Myerson, R. B. and M. A. Satterthwaite (1983). Efficient mechanisms for bilateral trading. *Journal of Economic Theory* 29(2), 265–281.

Napel, S. (1998). On the 'fairness' of the Nash bargaining solution. WIOR Discussion Paper 537, Universität Karlsruhe.

Napel, S. (1999a). Game theory and the evolution of fairness: A review article. *Homo Oeconomicus* 15(4), 581–586.

Napel, S. (1999b). The Holler-Packel axiomatization of the public good index completed. *Homo Oeconomicus* 15(4), 513–520.

Napel, S. (2000). Aspiration adaptation in the ultimatum minigame. WIOR Discussion Paper 596, University of Karlsruhe.

Napel, S. (2001). A note on the Holler-Packel axiomatization of the public good index (PGI). See Holler and Owen (2001), pp. 143–151.

Napel, S. and M. Widgrén (2000). Inferior players in simple games. ETLA Discussion Paper 734, Research Institute of the Finnish Economy.

Napel, S. and M. Widgrén (2001a). Inferior players in simple games. *International Journal of Game Theory* 30(2), 209–220.

Napel, S. and M. Widgrén (2001b). The power of an inferior player. In M. J. Holler, H. Kliemt, and D. Schmidtchen (Eds.), *Jahrbuch für Neue Politische Ökonomie*, Volume 19. Tübingen: Mohr Siebeck.

Nash, J. F. (1950a). The bargaining problem. *Econometrica* 18(2), 155–162.

Nash, J. F. (1950b). Equilibrium points in *n*-person games. *Proceedings of the National Academy of Sciences* 36, 48–49.

Nash, J. F. (1951). Non-cooperative games. *Annals of Mathematics* 54(2), 286–295.

Nash, J. F. (1953). Two-person cooperative bargaining games. *Econometrica* 21(1), 128–140.

Neelin, J., H. Sonnenschein, and M. Spiegel (1988). A further test of noncooperative bargaining theory: Comment. *American Economic Review* 78(4), 824–36.

Nurmi, H. (1998). *Rational Behaviour and Design of Institutions*. Cheltenham: Edward Elgar.

Nydegger, R. V. and G. Owen (1974). Two-person bargaining: An experimental test of Nash axioms. *International Journal of Game Theory* 3(4), 239–249.

Ochs, J. and A. E. Roth (1989). An experimental study of sequential bargaining. *American Economic Review 79*(3), 355–84.

Okada, A. (1996). A noncooperative coalitional bargaining game with random proposers. *Games and Economic Behavior 16*(1), 97–108.

Osborne, M. J. and A. Rubinstein (1990). *Bargaining and Markets*. San Diego, CA: Academic Press.

Osborne, M. J. and A. Rubinstein (1994). *A Course in Game Theory*. Cambridge, MA: MIT Press.

Ostmann, A. (1987). On the minimal representation of homogeneous games. *International Journal of Game Theory 16*(1), 69–81.

Owen, G. (1972). Multilinear extensions of games. *Management Science – Theory 18*(5), P64–P79.

Owen, G. (1975). Evaluation of a presidential election game. *American Political Science Review 69*, 947–953.

Owen, G. (1988). Multilinear extensions of games. See Roth (1988), pp. 139–151.

Owen, G. (1995). *Game Theory* (3rd ed.). San Diego, CA: Academic Press.

Pareto, V. (1909). *Manuel D'Économie Politique*. Genève: Librairie Droz.

Pazgal, A. (1997). Satisficing leads to cooperation in mutual interests games. *International Journal of Game Theory 26*(4), 439–453.

Penrose, L. S. (1946). The elementary statistics of majority voting. *Journal of the Royal Statistical Society 109*, 53–57.

Peters, R. (2000). Evolutionary stability in the ultimatum game. *Group Decision and Negotiation 9*(4), 325–324.

Ponsati, C. and J. Sakovics (1998). Rubinstein bargaining with two-sided outside options. *Economic Theory 11*(3), 667–672.

Poulsen, A. (2000). Commitment and information in ultimatum bargaining – An evolutionary analysis. Working paper, Department of Economics, University of Essex.

Prasnikar, V. and A. E. Roth (1992). Considerations of fairness and strategy: Experimental data from sequential games. *Quarterly Journal of Economics 107*(3), 865–888.

Rawls, J. (1971). *A Theory of Justice*. Cambridge, MA: Belknap Press.

Rawls, J. (1974). Some reasons for the maximin criterion. *American Economic Review, Papers and Proceedings 64*(2), 141–146.

Rawls, J. and E. Kelly (2001). *Justice As Fairness: A Restatement*. Cambridge, MA: Harvard University Press.

Riker, W. H. (1986). The first power index. *Social Choice and Welfare 3*, 293–295.

Robson, A. J. and F. Vega-Redondo (1996). Efficient equilibrium selection in evolutionary games with random matching. *Journal of Economic Theory 70*(1), 65–92.

Rosenmüller, J. (1981). *The Theory of Games and Markets*. Amsterdam: North-Holland.

Roth, A. E. (1979). *Axiomatic Models of Bargaining*, Volume 170 of *Lecture Notes in Economics and Mathematical Systems*. Berlin: Springer-Verlag.

Roth, A. E. (Ed.) (1988). *The Shapley Value – Essays in Honor of Lloyd S. Shapley*. Cambridge: Cambridge University Press.

Roth, A. E. (1995a). Bargaining experiments. See Kagel and Roth (1995), Chapter 4, pp. 253–348.

Roth, A. E. (1995b). Introduction to experimental economics. See Kagel and Roth (1995), Chapter 1, pp. 3–109.

Roth, A. E. and I. Erev (1995). Learning in extensive-form games: Experimental data and simple dynamic models of the intermediate term. *Games and Economic Behavior 8*(1), 164–212.

Roth, A. E. and M. W. K. Malouf (1979). Game-theoretic models and the role of information in bargaining. *Psychological Review 86*(6), 574–594.

Roth, A. E. and J. K. Murnighan (1982). The role of information in bargaining: An experimental study. *Econometrica 50*(2), 1123–1142.

Roth, A. E., V. Prasnikar, M. Okuno-Fujiwara, and S. Zamir (1991). Bargaining and market behavior in Jerusalem, Ljubljana, Pittsburgh, and Tokyo: An experimental study. *American Economic Review 81*(5), 1068–1095.

Roth, A. E. and U. G. Rothblum (1982). Risk aversion and Nash's solution for bargaining games with risky outcomes. *Econometrica 50*(3), 639–647.

Rubinstein, A. (1982). Perfect equilibrium in a bargaining model. *Econometrica 50*(1), 97–109.

Rubinstein, A. (1985a). A bargaining model with incomplete information about time preferences. *Econometrica 53*(5), 1151–1172.

Rubinstein, A. (1985b). Choice of conjectures in a bargaining game with incomplete information. In A. E. Roth (Ed.), *Game-Theoretic Models of Bargaining*, pp. 99–114. Cambridge: Cambridge University Press.

Rubinstein, A. (1987). A sequential strategic theory of bargaining. In T. F. Bewley (Ed.), *Advances in Economic Theory*, pp. 197–224. Cambridge: Cambridge University Press.

Sáez-Martí, M. and J. W. Weibull (1999). Clever agents in Young's evolutionary bargaining model. *Journal of Economic Theory 86*(2), 268–279.

Samuelson, L. (1997). *Evolutionary Games and Equilibrium Selection*. Cambridge, MA: MIT Press.

Sauermann, H. and R. Selten (1962). Anspruchsanpassungstheorie der Unternehmung. *Zeitschrift für die gesamte Staatswissenschaft 118*(4), 577–597.

Schelling, T. C. (1960). *The Strategy of Conflict*. Cambridge, MA: Harvard University Press.

Schlag, K. H. (1998). Justifying imitation. Mimeo, Economic Theory III, University of Bonn.

Selten, R. (1965). Spieltheoretische Behandlung eines Oligopolmodells mit Nachfrageträgheit. *Zeitschrift für die gesamte Staatswissenschaft 121*(2 and 4), 301–324 and 667–689.

Selten, R. (1975). Reexamination of the perfectness concept for equilibrium points in extensive games. *International Journal of Game Theory* 4(1), 25–55.

Selten, R. (1998). Aspiration adaptation theory. *Journal of Mathematical Psychology* 42, 191–214.

Sen, A. K. (1970). *Collective Choice and Social Welfare*. San Francisco, CA: Holden-Day.

Shaked, A. and J. Sutton (1984). Involuntary unemployment as a perfect equilibrium in a bargaining model. *Econometrica* 52(6), 1351–1364.

Shapley, L. S. (1953). A value for *n*-person games. In H. W. Kuhn and A. W. Tucker (Eds.), *Contributions to the theory of games*, Volume II, pp. 307–317. Princeton, NJ: Princeton University Press.

Shapley, L. S. and M. Shubik (1954). A method for evaluating the distribution of power in a committee system. *American Political Science Review* 48(3), 787–792.

Simon, C. P. and L. Blume (1994). *Mathematics for Economists*. New York: W. W. Norton.

Simon, H. A. (1955). A behavioral model of rational choice. *Quarterly Journal of Economics* 69(1), 99–118.

Simon, H. A. (1959). Theories of decision-making in economics and behavioral science. *American Economic Review* 49(3), 253–283.

Slembeck, T. (1999). Low information games – Experimental evidence on learning in ultimatum bargaining. Discussion Paper 9903, Department of Economics, University of St. Gallen.

Slonim, R. and A. E. Roth (1998). Learning in high stakes ultimatum games: An experiment in the Slovak Republic. *Econometrica* 66(3), 569–596.

Sobel, J. and I. Takahashi (1983). A multistage model of bargaining. *Review of Economic Studies* 50(3), 411–426.

Spiegel, H. W. (1991). *The Growth of Economic Thought* (3rd ed.). Durham: Duke University Press.

Ståhl, I. (1972). *Bargaining Theory*. Ph. D. thesis, Stockholm School of Economics.

Steunenberg, B., D. Schmidtchen, and C. Koboldt (1999). Strategic power in the European Union – Evaluating the distribution of power in policy games. *Journal of Theoretical Politics* 11(3), 339–366.

Stokey, N. and R. E. Lucas (1989). *Recursive Methods in Economic Dynamics*. Cambridge, MA: Harvard University Press.

Straffin, Jr., P. D. (1977). Homogeneity, independence and power indices. *Public Choice* 30, 107–118.

Straffin, Jr., P. D. (1988). The Shapley-Shubik and Banzhaf power indices as probabilities. See Roth (1988), pp. 71–81.

Straffin, Jr., P. D. (1994). Power and stability in politics. See Aumann and Hart (1994), Chapter 32, pp. 1127–1151.

Sugden, R. (2001). Ken Binmore's evolutionary social theory. *Economic Journal* 111(469), F213–F243.

Sutton, J. (1986). Non-cooperative bargaining theory: An introduction. *Review of Economic Studies* 53(5), 709–724.

Sydsæter, K., A. Strøm, and P. Berck (1999). *Economists' Mathematical Manual* (3rd ed.). Berlin: Springer.

Tarascio, V. J. (1992). A correction: On the geneology of the so-called Edgeworth-Bowley diagram. In M. Blaug (Ed.), *Alfred Marshall (1842-1924) and Francis Edgeworth (1845-1926)*, Pioneers in Economics Vol. 29, pp. 75–79. Aldershot: Edward Elgar.

Taylor, H. M. and S. Karlin (1998). *An Introduction to Stochastic Modeling* (3rd ed.). San Diego, CA: Academic Press.

Taylor, P. and L. Jonker (1978). Evolutionary stable strategies and game dynamics. *Mathematical Biosciences* 40(1-2), 145–156.

Thomson, W. (1994). Cooperative models of bargaining. See Aumann and Hart (1994), Chapter 35, pp. 1237–1284.

Tieman, A. F., H. Houba, and G. van der Laan (2000). On the level of cooperative behavior in a local interaction model. *Journal of Economics (Zeitschrift für Nationalökonomie)* 71(1), 1–30.

Van Bragt, D. D. B., E. H. Gerding, and J. A. La Poutré (2000). Equilibrium selection in alternating-offers bargaining models – The evolutionary computing approach. Report SEN-R0013, Stichting Mathematisch Centrum, Amsterdam.

Van Damme, E. (1991). *Stability and Perfection of Nash Equilibria* (2nd ed.). Berlin: Springer-Verlag.

Van Damme, E., R. Selten, and E. Winter (1990). Alternating bid bargaining with a smallest money unit. *Games and Economic Behavior* 2(2), 188–201.

Varian, H. R. (1974). Equity, envy, and efficiency. *Journal of Economic Theory* 9, 63–91.

Von Neumann, J. and O. Morgenstern (1953). *Theory of Games and Economic Behavior* (3rd ed.). Princeton, NJ: Princeton University Press. (1st edition, 1944).

Weibull, J. W. (1995). *Evolutionary Game Theory*. Cambridge, MA: MIT Press.

Weibull, J. W. (2000). Testing game theory. Working Paper 382, Stockholm School of Economics. (Revised: January 2001).

Widgrén, M. and S. Napel (2001). The power of a spatially inferior player. ETLA Discussion Paper 771, Research Institute of the Finnish Economy. Forthcoming in *Homo oeconomicus*.

Winter, S. G. (1971). Satisficing, selection, and the innovating remnant. *Quarterly Journal of Economics* 85(2), 237–261.

Yang, C.-L., J. Weimann, and A. Mitropoulos (1999). An experiment on bargaining power in simple sequential games. FEMM working paper, Universität Magdeburg. (Revised: Game structure and bargaining in sequential mini-games: An experiment; April 2001).

Young, H. P. (1993a). The evolution of conventions. *Econometrica 61*(1), 57–84.

Young, H. P. (1993b). An evolutionary model of bargaining. *Journal of Economic Theory 59*(1), 145–168.

Young, H. P. (1998). *Individual Strategy and Social Structure*. Princeton, NJ: Princeton University Press.

Zeuthen, F. (1930). *Problems of Monopoly and Economic Warfare*. London: George Routledge & Sons. Reprinted by Routledge & Kegan Paul, London, 1967.

Index

Druck: Strauss Offsetdruck, Mörlenbach
Verarbeitung: Schäffer, Grünstadt